Essentials of Geology

FOURTH EDITION

Essentials of Geology

FOURTH EDITION

Stephen Marshak

UNIVERSITY OF ILLINOIS

W. W. NORTON & COMPANY

NEW YORK   LONDON

W. W. Norton & Company has been independent since its founding in 1923, when William Warder Norton and Mary D. Herter Norton first published lectures delivered at the People's Institute, the adult education division of New York City's Cooper Union. The firm soon expanded its program beyond the Institute, publishing books by celebrated academics from America and abroad. By mid-century, the two major pillars of Norton's publishing program—trade books and college texts—were firmly established. In the 1950s, the Norton family transferred control of the company to its employees, and today—with a staff of four hundred and a comparable number of trade, college, and professional titles published each year—W. W. Norton & Company stands as the largest and oldest publishing house owned wholly by its employees.

Illustrations by Precision Graphics
Composition by CodeMantra, Inc.
Manufacturing by Courier—Kendallville, IN

The text of this book is set in Adobe Caslon, with display in Conduit, Din, Marbrook BQ, and Univers.

Editor: Eric Svendsen
Senior project editors: Thomas Foley and Lory Frenkel
Production manager: Benjamin Reynolds
Copy editor: Jennifer Harris
Managing editor, College: Marian Johnson
Book design: Lissi Sigillo
Art director: Rubina Yeh
Media editor: Rob Bellinger
Digital media editorial assistant, sciences: Paula Iborra
Associate supplements editor: Callinda Taylor
Marketing manager, physical sciences: Stacy Loyal
Photography director: Trish Marx
Editorial assistants: Hannah Bachman and Alicia González-Gross

Library of Congress Cataloging-in-Publication Data

Marshak, Stephen, 1955-
   Essentials of geology / Stephen Marshak. — 4th ed.
      p. cm.
   Includes index.
      ISBN 978-0-393-91939-4 (pbk.)
   1. Geology—Textbooks. I. Title.
   QE28.M3415 2013
   551—dc23
                                 2012037689

W. W. Norton & Company, Inc., 500 Fifth Avenue, New York, NY 10110
wwnorton.com
W. W. Norton & Company Ltd., Castle House, 75/76 Wells Street, London W1T 3QT

1 2 3 4 5 6 7 8 9 0

Cover photo: Wave-carved granite cliffs along the Côte Sauvage ("Wild Coast"), on the south side of Brittany, France (Lat 47°30'36.11"N, Long 3°9'1.17"W)

DEDICATION

To Kathy, David, Emma, and Michelle

Brief Contents

Contents

CHAPTER 3

Patterns in Nature: Minerals • 71

INTERLUDE A

Rock Groups • 88

CHAPTER 4

Up from the Inferno: Magma and Igneous Rocks • 97

INTERLUDE F

An Introduction to Landscapes and the Hydrologic Cycle • 386

CHAPTER 13

Unsafe Ground: Landslides and Other Mass Movements • 397

Preface

Narrative Themes

Why do earthquakes, volcanoes, floods, and landslides happen? What causes mountains to rise? How do beautiful landscapes develop? Do climate and life change through time? When did the Earth form and by what process? Where do we dig to find valuable metals and where do we drill to find oil? Does sea level change? Can continents move? The study of geology addresses these important questions and many more. But from the birth of the discipline in the late 18th century until the mid-20th century, geologists considered each question largely in isolation, without pondering its relation to the others. This approach changed, beginning in the 1960s, in response to the formulation of two "paradigm-shifting" ideas that have unified thinking about the Earth and its features. The first idea, called the *theory of plate tectonics*, states that the Earth's outer shell consists of discrete plates that slowly move relative to each other so that the map of our planet continuously changes. Plate interactions cause earthquakes and volcanoes, build mountains, provide gases that make up the atmosphere, and affect the distribution of life on Earth. The second idea, called the *Earth System concept*, emphasizes that our planet's water, land, atmosphere, and living inhabitants are dynamically interconnected. In the Earth System, materials constantly cycle among various living and nonliving reservoirs on, above, and within the planet. Thus, we have come to realize that the history of life is intimately linked to the history of the physical Earth.

Essentials of Geology, Fourth Edition, is an introduction to the study of our planet that employs the theory of plate tectonics and the concept of the Earth System throughout to weave together a number of narrative themes, including the following:

1. The solid Earth, the oceans, the atmosphere, and life interact in complex ways, yielding a planet that is unique in the Solar System.

2. Most geologic processes reflect the interactions among plates.

3. The Earth is a planet, formed like other planets from dust and gas. But, in contrast to other planets, the Earth is a dynamic place on which new geologic features continue to form and old ones continue to be destroyed.

4. The Earth is very old—about 4.57 billion years have passed since its birth. During this time, the surface, subsurface, and atmosphere of the planet have changed, and life has evolved.

5. Internal processes (driven by Earth's internal heat) and external processes (driven by heat from the Sun) interact at the Earth's surface to produce complex landscapes.

6. Geologic knowledge can help society understand natural hazards such as earthquakes, volcanoes, landslides, and floods, and in some cases can reduce the danger that these hazards pose.

7. Energy and mineral resources come from the Earth and are formed by geologic phenomena. Geologic study can help locate these resources and mitigate the consequences of their use.

8. Physical features of the Earth are linked to life processes, and vice versa.

9. Science comes from observation; people make scientific discoveries.

10. Geology utilizes ideas from physics, chemistry, and biology, so the study of geology provides an excellent means to improve science literacy.

These narrative themes serve as the *take-home message* of this book, a message that students should remember long after they finish their introductory geology course. In effect, they provide a mental framework on which students can organize and connect ideas, and develop a modern, coherent image of our planet.

Pedagogical Approach

Students learn best from textbooks when they can actively engage with a combination of narrative text and narrative art. Some students respond more to the words, which help them to organize information, provide answers to questions, fill in the essential steps that link ideas together, and develop a personal context for understanding information. Some students respond more to narrative art—art designed to tell a story—for visual images help students comprehend and remember processes. And some respond to question-and-answer-based active learning, an approach where students can in effect "practice" their knowledge in real time. *Essentials of Geology*, Fourth Edition provides all three of these learning tools. The text has been crafted to be engaging and to carry students forward in a narrative form, the art has been configured to tell a story, the chapters are laid out to help students internalize key principles, and the online activities have been designed to both engage students and provide active feedback. For example, *Did You Ever Wonder* panels prompt students to connect new information to their existing knowledge base by asking geology-related questions that they have probably already thought about. *Take-Home Message* panels at the end of each section help students solidify key themes before proceeding to the next section. Questions at the end of each chapter not only test basic knowledge, but also stimulate critical thinking. New SmartWork online homework helps students prepare with automatic feedback, visual drag-and-drop labeling, and "hot spot" reviews. Finally, the *See for Yourself* and *Geotour* features guide students on virtual field trips, via *Google Earth*™, to locales around the globe where they can apply their newly acquired knowledge to the interpretation of real-world geologic features.

Organization

The topics covered in this book have been arranged so that students can build their knowledge of geology on a foundation of overarching principles. Thus, the book starts by considering how the Earth formed, and how it is structured, overall, from its surface to its center. With this basic background, students can delve into plate tectonics, the grand unifying theory of geology. Plate tectonics appears early in the book, so that students can use the theory as a foundation from which they can interpret and link ideas presented in subsequent chapters. Knowledge of plate tectonics, for example, helps students understand the suite of chapters on minerals, rocks, and the rock cycle. Knowledge of plate tectonics and rocks together, in turn, provides a basis for studying volcanoes, earthquakes, and mountains. And with this background, students are prepared to see how the map of the Earth has changed through the vast expanse of geologic time, and how energy and mineral resources have developed. The book's final chapters address processes and problems occurring at or near the Earth's surface, from the unstable slopes of hills, down the course of rivers, to the shores of the sea and beyond. This section concludes with a topic of growing concern in society—global change, particularly climate change.

Although the sequence of chapters was chosen for a reason, this book is designed to be flexible so that instructors can choose their own strategies for teaching geology. Thus, each chapter is self-contained, and we reiterate relevant material where necessary.

Special and Updated Features of This Edition

Narrative Art and *What a Geologist Sees*

To help students visualize topics, this book is lavishly illustrated, with figures designed to provide a realistic context for interpreting geologic features without overwhelming students with extraneous detail. In this edition, many drawings and photographs have been integrated into *narrative art* that has been laid out, labeled, and annotated to tell a story—the figures are drawn to teach! Subcaptions are positioned adjacent to the relevant parts of a figure, labels point out key features, and balloons provide important detail. Subparts have been arranged to convey time progression, where relevant. Color schemes in drawings have been tied to those of relevant photos, so that students can easily visualize the relationships between drawings and photos. In some examples, photographs are accompanied by annotated sketches labeled *What a Geologist Sees*, which help students to be certain that they actually see the specific features that the photo was intended to show.

Featured Paintings: *Geology at a Glance*

In addition to individual figures, renowned British artist Gary Hincks has created spectacular two-page annotated paintings, called *Geology at a Glance*. These paintings integrate key concepts introduced in the chapters and visually emphasize the relationships among components of the Earth System. And, they provide students with a way to review a subject . . . at a glance.

New Coverage of Current Topics

To ensure that *Essentials of Geology*, Fourth Edition, reflects the latest research discoveries and helps students understand the geologic events that have been featured in news headlines, we have updated many topics throughout the book. For example, this edition provides insightful treatment of catastrophic earthquakes in Haiti, Japan, and New Zealand; illustrates recent massive tornado outbreaks; explains the significance of the nonconventional gas reserves of the Marcellus Shale; and characterizes the critical sustainability issues related to our increasing reliance on rare earth elements.

See for Yourself — Using *Google Earth*™

Visiting Field Sites Identified in the Text

There's no better way to appreciate geology than to see it first-hand in the field. The challenge is that the great variety of geologic features that we discuss in this book can't be visited from any one locality. So even if your class can take geology field trips during the semester, you'll at most see examples of just a few geologic settings. Fortunately, *Google Earth*™ makes it possible for you to fly to spectacular geologic field sites anywhere in the world in a matter of seconds—you can take a virtual field trip electronically. At the end of each chapter in this book, you will find a *See For Yourself* section identifying geologic sites that you can explore on your own personal computer (Mac or PC) using *Google Earth*™ software, or on your Apple/Android smartphone or tablet with the appropriate *Google Earth*™ app.

To get started, follow these three simple steps:

1 Check to see whether *Google Earth*™ is installed on your personal computer, smartphone, or tablet. If not, please download the software from **earth.google.com** or the app from the Apple or Android app store.

2 In the *See For Yourself* section at the end of a chapter, find a site that you're interested in visiting. In addition to a thumbnail photo and very brief description of the site (highlighting what you will see at the site), we provide the latitude and longitude of the site.

3 Open *Google Earth*™, and enter the coordinates of the site in the search window. As an example, let's find Mt. Fuji, a beautiful volcano in Japan. We specify the coordinates in the book as follows:

Latitude   35°21′41.78″N,
Longitude   138°43′50.74″E

Type these coordinates into the search window as:

35 21 41.78N, 138 43 50.74E

Note that the degree (°), minute (′), and second (″) symbols are simply left as blank spaces.

When you click Enter or Return, your device will bring you to the viewpoint right above Mt. Fuji illustrated by the thumbnail on the left. Note that you can use the tools built into *Google Earth*™ to vary the elevation, tilt, orientation, and position of your viewpoint. The thumbnail on the right shows the view you'll see of the same location if you tilt your viewing direction and look north.

View looking down.   View looking north.

Need More Help? If you aren't up and running, please visit wwnorton.com/web/egeo4/see to find a video showing you how to download and install *Google Earth*™, more detailed instructions on how to find the *See for Yourself* sites, additional sites not listed in this book, links to *Google Earth*™ videos describing basic functions, and links to any hardware and software requirements. Also, notes addressing important *Google Earth*™ updates will be available at this site.

We also offer a separate book—the *Geotours Workbook* (ISBN 978-0-393-91891-5)—that identifies even more interesting geologic sites to visit, provides active-learning exercises linked to the sites, and explains how you can create your own virtual field trips.

New SmartWork online Tutorial and Assessment System for Geology

The SmartWork online tutorial and assessment system available to users of *Essentials of Geology*, Fourth Edition features visual assignments that provide students with answer-specific feedback. Students get the coaching they need to work through the assignments, while instructors get real-time assessment of student progress with automatic grading and item analysis. Image-based drag-and-drop and hot-spot questions make use of carefully developed images, and additional *What a Geologist Sees* figures have been created exclusively for SmartWork. Additional video, animation, and conceptual questions challenge students to apply their understanding of important concepts identified by the author in each chapter's *Take-Home Message*. SmartWork also provides Reading Quizzes, *Geotour*-guided inquiry activities using *Google Earth*™, and approximately 1,000 basic content questions.

Features Designed to Help Students Prepare and Review

Each chapter begins with chapter objectives that frame the major concepts for students, and every section ends with a *Take-Home Message*, a brief summary that helps students identify and remember the highlight of the section before moving on to the next. *Did You Ever Wonder* questions are sprinkled within the chapter to both engage students and answer common questions about geology. Each chapter now concludes with a integrated two-page *Chapter Review*, designed to pull together summary points, key terms, basic review questions, *On Further Thought* critical-thinking questions, and *See for Yourself* sites into a single, visually compact, feature. The *See for Yourself* sites introduce students to field examples of geology accessible by using *Google Earth*™.

Interludes

The book contains several Interludes. These "mini-chapters" focus on key topics that are self-contained but are not broad enough to require an entire chapter. Placing some content in the Interludes not only keeps chapters reasonable in length, but also provides additional flexibility in sequencing topics within a course.

Societal Issues

We address geology's practical applications in several chapters, providing students with an opportunity to learn about energy resources, mineral resources, global change, and natural disasters. *Science and Society* features, provided for free on the open student website and in Norton's LMS coursepacks, challenge students to apply material that they learn in *Essentials of Geology*, Fourth Edition to the interpretation of news articles and publicly available geologic data.

Supplementary Materials

 New SmartWork Online Tutorial and Assessment System. The new SmartWork online assessment available for use with *Essentials of Geology*, Fourth Edition features visual assignments with focused feedback. Because students learn best when they can interact with art as well as with text, SmartWork includes drag-and-drop figure-based questions, animation- and video-based questions, and *What a Geologist Sees* photo interpretations. SmartWork also provides questions based on real field examples, via the *Geotours Workbook*, and helps students check their knowledge as they go by working with reading-based questions. Designed to be intuitive and easy to use (for both students and instructors), SmartWork makes it a snap to assign, assess, and report on student performance, and keep the class on track.

Student Access Codes. If students need to purchase an access code for SmartWork, they can order through their college bookstore using ISBN 978–0–393–91939–4 for Smart-Work with e-book. Immediate online access can also be purchased at smartwork.wwnorton.com—select the option to buy a registration code before you create your account. Instructors can request their own SmartWork course at wwnorton.com/instructors.

Art Files and PowerPoints

> *Enhanced Art PowerPoints*—Designed for instant classroom use, these slides utilize photographs and line art from the book in a form that has been optimized for use in the PowerPoint environment. The art has been relabeled and resized for projection formats. *Enhanced Art PowerPoints* also include supplemental photographs.

> *Lecture Bullet PowerPoints*—These slides include both art and bulleted text for direct use either in lectures or as student handouts.

> *Labeled and Unlabeled Art PowerPoints*—These include all art from the book formatted as JPEGs that have been prepasted into PowerPoints. We offer one set in which all labeling has been stripped and one set in which labeling remains.

> *Art JPEGs*—We provide a complete file of individual JPEGs for art and photographs used in the book.

> *Monthly Update PowerPoints*—W. W. Norton & Company offers a monthly update service that provides new Power-Point slides, with instructor support, covering recent geologic events. Starting in December 2012, these updates will help instructors keep their classes current.

New Animations and Videos. The book's accompanying Instructor Resource DVD provides a rich collection of animations to illustrate geologic processes. The set includes

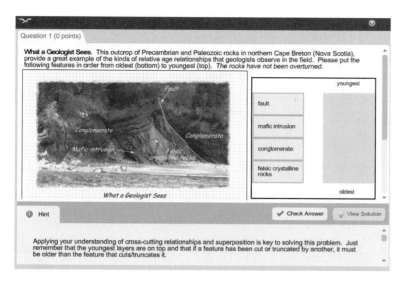

SmartWork

The new SmartWork online assessment system available for use with Essentials of Geology features highly visual questions with immediate, answer-specific feedback.

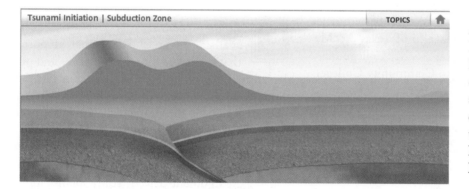

Animations

New interactive animations help instructors to demonstrate concepts and then allow students to explore those concepts on their own. New topics include Tsunami Initiation, Faults, and Metamorphic Change.

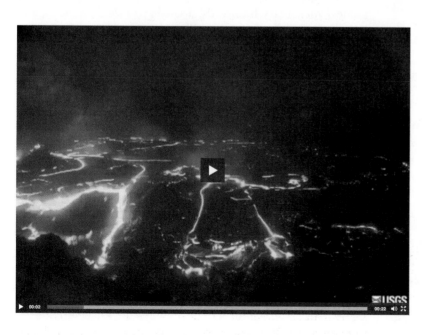

Videos

A new real-world video collection makes clips easily available in a reliable and convenient format. In SmartWork, questions with answer-specific feedback help students connect the videos to what they've learned in class.

10 new animations, developed by Alex Glass of Duke University, that allow you to control variables. In addition, working with Melissa Hudley of the University of North Carolina, Chapel Hill, Heather Lehto of Angelo State University, and Meghan Lindsey of the University of South Florida, our DVD, coursepacks and instructor support website contain over 50 streaming videos of geologic processes including content developed by IRIS. All animations and videos are ready to go and perfect for classroom or online use.

Norton Media Library Instructor's DVD-ROM. The Instructor's DVD offers a wealth of easy-to-use multimedia resources all structured around the text. Resources include all of the PowerPoints art files, animations, and videos described earlier in the preface, and electronic versions of the Instructor's Manual, Test Bank, and Exam View test generation software. Further resources include GeoQuiz clicker questions, and supporting files for using *Google Earth*™.

Instructor's Manual and Test Bank. The Instructor's Manual and Test Bank, prepared by John Werner of Seminole State College of Florida; Jacalyn Gorczynski of Texas A&M, Corpus Christi; Haley Wasson of Texas A&M, College Station; and Daniel Wynne of Sacramento Community College, are designed to help instructors prepare lectures and exams. The Instructor's Manual contains detailed Learning Objectives, Chapter Summaries, and complete answers to end-of-chapter review and *On Further Thought* questions. The Test Bank has been developed using the Norton Assessment Guidelines, and each chapter of the Test Bank consists of three types of questions, classified according to Norton's taxonomy of knowledge types and by section and difficulty, making it easy to construct tests and quizzes that are meaningful and diagnostic. Test questions were also reviewed by instructors (we would like to thank Judy McIlrath of the University of South Florida, Marek Cichanski of DeAnza College, and Karen Koy of Missouri Western State University for their feedback). These supplements are available in print paperback and on DVD, and are also downloadable from wwnorton.com/instructors.

Instructor's Website—wwnorton.com/instructors. Online access is available to a rich array of resources: Test Bank, Instructor's Manual, PowerPoints, JPEGs, *Google Earth*™ file of sites from the text, art from the text, and WebCT- and Blackboard-ready content.

Coursepacks. Available at no cost to professors or students, Norton *Coursepacks* bring high-quality Norton digital media into a new or existing online course. For *Essentials of Geology*, Fourth Edition, content includes Test Bank, Reading Quizzes, Quiz+ questions, *Geotour* questions, Guides to Reading, animations, streaming video, and links to the e-book.

The *Google Earth*™ *Geotours Workbook*. Created by Scott Wilkerson, Beth Wilkerson, and Stephen Marshak, *Geotours* are active-learning opportunities that take students on virtual field trips to see outstanding examples of geology at localities around the world. Arranged by topic, questions for the *Geotours* have been designed for auto-grading, and are available as Worksheets both in print format (these come free with the book and include complete user instructions and advanced instruction) and electronically with auto-grading through SmartWork or your campus LMS. Request a sample copy to preview each worksheet.

See for Yourself *Google Earth*™ Sample Site File. Users who simply want access to sample field sites for classroom presentations or distribution to students can download the sites from the book's *See for Yourself* review sections. This single download is available at the Norton instructor download site as well as from wwnorton.com/studyspace.

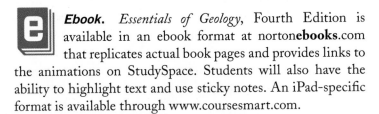

Ebook. *Essentials of Geology*, Fourth Edition is available in an ebook format at norton**ebooks**.com that replicates actual book pages and provides links to the animations on StudySpace. Students will also have the ability to highlight text and use sticky notes. An iPad-specific format is available through www.coursesmart.com.

StudySpace—wwnorton.com/studyspace. Free and open for students, StudySpace offers students assignment-driven study plans for each chapter. Materials include Quiz+ diagnostic quizzes, *Science and Society* features that challenge students to use course concepts in analyzing news articles and real-time geologic data, reading guides, vocabulary flashcards, a *Geology in the News* feed, and information on how to best utilize the *Google Earth*™ materials provided for this book. StudySpace includes a video designed to help with start-up, all the *See For Yourself* sites in one downloadable file, instructions to enter latitude and longitude coordinates for each site, and links to *Google Earth*™ tutorials and information on updates.

Acknowledgments

I am very grateful for the assistance of many people in bringing this book from the concept stage to the shelf in the first place, and for helping to provide the momentum needed to make this revision take shape.

First and foremost, I wish to thank my wife, Kathy, who served as coauthor and in-home project manager for this book. Kathy helped to construct the manuscript by incorporating changes added to this book's parent, *Earth: Portrait of a Planet*, managed the manuscript traffic between our household and the publisher, cross-checked proofs, maintained the in-home art list and style sheet, and served as an invaluable extra set of eyes for catching errors. This book would not have happened without Kathy. I also wish to thank my daughter, Emma, and son, David, for their willingness to adopt "the book" as a member of the household when they were growing up, and to endure the overabundance of geo-photo stops on family trips. Emma also helped develop the concept of narrative art used in the book, and provided feedback about how the book works from a student's perspective.

I am very grateful to all the staff of W. W. Norton & Company for their incredible efforts during the development of my books over the past two decades. It has been a privilege to work with a company that is willing to work so closely with its authors. In particular, I would like to thank Thom Foley, the project manager for the book, who did a Herculean job of overseeing what proved to be an unexpectedly complicated process of managing the chapter proofs and the figures, all while remaining incredibly calm. Thom invested untold hours in sorting out composition and design issues—it's thanks to Thom that this edition made it to the shelf on schedule. Many thanks to the senior editor, Eric Svendsen, who injected new enthusiasm and ideas into the project. Eric's experience and skill have guided the book in new directions and have connected the project to new trends in science pedagogy and book design. I also greatly appreciate the efforts of Rob Bellinger for his innovative approach to ancillary development and for overseeing the development of the SmartWork supplements; Trish Marx for her expert and thoughtful editing of the photo collection and permissions; Ben Reynolds for coordinating the back-and-forth between the publisher and various suppliers; Stacy Loyal, who has so ably assumed the mantle of sales manager for the book; Callinda Taylor for capably handling the ancillaries; Paula Iborra for assisting Rob with everything emedia; and Jennifer Harris for her excellent copyediting work. *Essentials of Geology* would not have reached a fourth edition without the inspiration and support of Jack Repcheck, the editor of the previous three editions. Jack proposed several of the features that attracted readers to the book in the first place, and continues to provide guidance and support. I also wish to thank Susan Gaustad, the outstanding developmental editor of the first edition, who helped refine the prose style of the book.

Production of the illustrations has involved many people over many years. I am indebted to the staff of *Precision Graphics*, who have helped to create the style of the figures and have accommodated countless changes and tweaks without complaint. Stan Maddock and Becky Oles have been at the core of this effort, and I am forever grateful for their talent and hard work. Kristina Seymour has done a wonderful job of project management over the past several years, and is always a pleasure to work with. Jon Prince and Jeff Griffin creatively programmed the animations under the careful supervision of Andrew Troutt.

It has been great fun to interact with Gary Hincks, who painted the incredible two-page spreads, in part using his own designs and geologic insights. Some of Gary's paintings originally appeared in *Earth Story* (BBC Worldwide, 1998) and were based on illustrations conceived with Simon Lamb and Felicity Maxwell. Others were developed specifically for *Earth: Portrait of a Planet* and *Essentials of Geology*. Some of the chapter-opening quotes were found in *Language of the Earth*, compiled by F. T. Rhodes and R. O. Stone (Pergamon, 1981).

The four editions of this book and of its parent, *Earth: Portrait of a Planet*, have benefited greatly from input by expert reviewers for specific chapters, by general reviewers of the entire book, and by comments from faculty and students who have used the book and were kind enough to contact me or the publisher. In particular, in this edition I would like to thank Michael Rygel of SUNY Potsdam, who proofed every chapter as we created the initial book pages. Judy McIlrath of the University of South Florida, Marek Cichanski of DeAnza College, and Karen Koy of Missouri Western State University provided valuable feedback on the Test Bank questions and Kurt Wilkie of Washington State University has offered extremely helpful input on both the book and the coursepack. Other reviewers who have provided helpful feedback for this and previous editions include:

Jack C. Allen, *Bucknell University*
David W. Anderson, *San Jose State University*
Martin Appold, *University of Missouri, Columbia*
Philip Astwood, *University of South Carolina*
Eric Baer, *Highline University*
Victor Baker, *University of Arizona*
Julie Baldwin, *University of Montana*
Sandra Barr, *Acadia University*
Keith Bell, *Carleton University*
Mary Lou Bevier, *University of British Columbia*
Jim Black, *Tarrant County College*
Daniel Blake, *University of Illinois*
Ted Bornhorst, *Michigan Technological University*
Michael Bradley, *Eastern Michigan University*
Mike Branney, *University of Leicester, UK*
Sam Browning, *Massachusetts Institute of Technology*
Bill Buhay, *University of Winnipeg*
Rachel Burks, *Towson University*
Peter Burns, *University of Notre Dame*
Katherine Cashman, *University of Oregon*

George S. Clark, *University of Manitoba*
Kevin Cole, *Grand Valley State University*
Patrick M. Colgan, *Northeastern University*
Peter Copeland, *University of Houston*
John W. Creasy, *Bates College*
Norbert Cygan, *Chevron Oil, retired*
Michael Dalman, *Blinn College*
Peter DeCelles, *University of Arizona*
Carlos Dengo, *ExxonMobil Exploration Company*
John Dewey, *University of California, Davis*
Charles Dimmick, *Central Connecticut State University*
Robert T. Dodd, *Stony Brook University*
Missy Eppes, *University of North Carolina, Charlotte*
Eric Essene, *University of Michigan*
James E. Evans, *Bowling Green State University*
Susan Everett, *University of Michigan, Dearborn*
Dori Farthing, *State University of New York, Geneseo*
Grant Ferguson, *St. Francis Xavier University*
Eric Ferré, *Southern Illinois University*
Leon Follmer, *Illinois Geological Survey*
Nels Forman, *University of North Dakota*
Bruce Fouke, *University of Illinois*
David Furbish, *Vanderbilt University*
Steve Gao, *University of Missouri*
Grant Garvin, *John Hopkins University*
Christopher Geiss, *Trinity College, Connecticut*
Gayle Gleason, *State University of New York, Cortland*
Cyrena Goodrich, *Kingsborough Community College*
William D. Gosnold, *University of North Dakota*
Lisa Greer, *William & Mary College*
Steve Guggenheim, *University of Illinois, Chicago*
Henry Halls, *University of Toronto, Mississuaga*
Bryce M. Hand, *Syracuse University*
Anders Hellstrom, *Stockholm University*
Tom Henyey, *University of South Carolina*
James Hinthorne, *University of Texas, Pan American*
Paul Hoffman, *Harvard University*
Curtis Hollabaugh, *University of West Georgia*
Bernie Housen, *Western Washington University*
Mary Hubbard, *Kansas State University*
Paul Hudak, *University of North Texas*
Warren Huff, *University of Cincinnati*
Neal Iverson, *Iowa State University*
Charles Jones, *University of Pittsburgh*
Donna M. Jurdy, *Northwestern University*

Thomas Juster, *University of Southern Florida*
H. Karlsson, *Texas Tech*
Daniel Karner, *Sonoma State University*
Dennis Kent, *Lamont Doherty/Rutgers*
Charles Kerton, *Iowa State University*
Susan Kieffer, *University of Illinois*
Jeffrey Knott, *California State University, Fullerton*
Ulrich Kruse, *University of Illinois*
Robert S. Kuhlman, *Montgomery County Community College*
Lee Kump, *Pennsylvania State University*
David R. Lageson, *Montana State University*
Robert Lawrence, *Oregon State University*
Scott Lockert, *Bluefield Holdings*
Leland Timothy Long, *Georgia Tech*
Craig Lundstrom, *University of Illinois*
John A. Madsen, *University of Delaware*
Jerry Magloughlin, *Colorado State University*
Jennifer McGuire, *Texas A&M University*
Judy McIlrath, *University of South Florida*
Paul Meijer, *Utrecht University, Netherlands*
Jamie Dustin Mitchem, *California University of Pennsylvania*
Alan Mix, *Oregon State University*
Otto Muller, *Alfred University*
Kathy Nagy, *University of Illinois, Chicago*
Pamela Nelson, *Glendale Community College*
Robert Nowack, *Purdue University*
Charlie Onasch, *Bowling Green State University*
David Osleger, *University of California, Davis*
Eric Peterson, *Illinois State University*
Ginny Peterson, *Grand Valley State University*
Stephen Piercey, *Laurentian University*
Adrian Pittari, *University of Waikato, New Zealand*
Lisa M. Pratt, *Indiana University*
Mark Ragan, *University of Iowa*
Robert Rauber, *University of Illinois*
Bob Reynolds, *Central Oregon Community College*
Joshua J. Roering, *University of Oregon*
Eric Sandvol, *University of Missouri*
William E. Sanford, *Colorado State University*
Jeffrey Schaffer, *Napa Valley Community College*
Roy Schlische, *Rutgers University*
Sahlemedhin Sertsu, *Bowie State University*
Doug Shakel, *Pima Community College*
Anne Sheehan, *University of Colorado*
Roger D. Shew, *University of North Carolina, Wilmington*

Norma Small-Warren, *Howard University*

Donny Smoak, *University of South Florida*

David Sparks, *Texas A&M University*

Angela Speck, *University of Missouri*

Tim Stark, *University of Illinois*

Seth Stein, *Northwestern University*

David Stetty, *Jacksonville State University*

Kevin G. Stewart, *University of North Carolina, Chapel Hill*

Michael Stewart, *University of Illinois*

Don Stierman, *University of Toledo*

Gina Marie Seegers Szablewski, *University of Wisconsin, Milwaukee*

Barbara Tewksbury, *Hamilton College*

Thomas M. Tharp, *Purdue University*

Kathryn Thornbjarnarson, *San Diego State University*

Basil Tikoff, *University of Wisconsin*

Spencer Titley, *University of Arizona*

Robert T. Todd, *Stony Brook University*

Torbjörn Törnqvist, *University of Illinois, Chicago*

Jon Tso, *Radford University*

Stacey Verardo, *George Mason University*

Barry Weaver, *University of Oklahoma*

John Werner, *Seminole State College of Florida*

Alan Whittington, *University of Missouri*

John Wickham, *University of Texas, Arlington*

Lorraine Wolf, *Auburn University*

Christopher J. Woltemade, *Shippensburg University*

I apologize if I inadvertently left anyone off this list.

About the Author

Stephen Marshak is a professor of geology at the University of Illinois, Urbana-Champaign, where he is also the director of the School of Earth, Society, and Environment. He holds an A.B. from Cornell University, an M.S. from the University of Arizona, and a Ph.D. from Columbia University. Steve's research interests in structural geology and tectonics have taken him into the field on several continents. He loves teaching and has won his college's and university's highest teaching awards, as well as the Neil Miner Award of the National Association of Geoscience Teachers "for exceptional contributions to the stimulation of interest in the earth sciences." In addition to *Essentials of Geology*, Steve has authored *Earth: Portrait of a Planet* and has co-authored *Laboratory Manual for Introductory Geology*; *Earth Structure: An Introduction to Structural Geology and Tectonics*; and *Basic Methods of Structural Geology*.

Thanks!

I am very grateful to the students who engaged so energetically with earlier editions of this book, and to the instructors who have selected this book for their classes. I welcome your comments and corrections and can be reached at smarshak@illinois.edu.

Stephen Marshak

Geology, perhaps more than any other department of natural philosophy, is a science of contemplation. It demands only an enquiring mind and senses alive to the facts almost everywhere presented in nature.

—Sir Humphry Davy (British scientist, 1778–1829)

And Just What Is Geology?

Students can see the Earth System at a glance on a sea cliff along the coast of Ireland. Here, sunlight, air, water, rock, and life all interact to produce a complex and fascinating landscape.

Civilization exists by geological consent, subject to change without notice.

—Will Durant (1885–1981)

P.1 In Search of Ideas

Our C-130 Hercules transport plane rose from a smooth ice runway on the frozen sea surface at McMurdo Station, Antarctica, and headed south to spend a month studying unusual rocks exposed on a cliff about 250 km away. We climbed past the smoking summit of Mt. Erebus, Earth's southernmost volcano, and for the next hour flew along the Transantarctic Mountains, a ridge of rock that divides the continent into two parts, East Antarctica and West Antarctica (**Fig. P.1**). Glaciers—sheets or rivers of ice that last all year—cover almost all of Antarctica. To the right of the plane we could see a continental glacier, a vast sheet of ice thousands of kilometers across and up to 4.7 km (15,000 ft) thick, that covers East Antarctica. The surface of this ice sheet forms a frigid high plain called the Polar Plateau. To the left we could see numerous valley glaciers, rivers of ice, that slowly carry ice from the Polar Plateau through gaps in the Transantarctic Mountains down to the frozen Ross Sea.

Suddenly, we heard the engines slow. As the plane descended, it lowered its ski-equipped landing gear. The loadmaster shouted an abbreviated reminder of the emergency alarm code: "If you hear three short blasts of the siren, hold on for dear life!" The plane touched the surface of our first choice for a landing spot, the ice at the base of the rock cliff we wanted to study. *Wham, wham, wham, wham!!!!* As the skis slammed into frozen snowdrifts on the ice surface at about 180 km an hour, it seemed as though a fairy-tale giant was shaking the plane. Seconds later, the landing aborted, we were airborne again, looking for a softer runway above the cliff. Finally, we landed in a field of deep snow, unloaded, and bade farewell to the plane. When the plane passed beyond the horizon, the silence of Antarctica hit us—no trees rustled, no dogs barked, and no traffic rumbled in this stark land of black rock and white ice. It would take us a day and a half to haul our sleds of food and equipment down to our study site (see Fig. P.1). All this to look at a few dumb rocks?

Geologists—scientists who study the Earth—explore remote regions like Antarctica almost routinely. Such efforts often strike people in other professions as a strange way to make a living, as implied by Scottish poet Walter Scott's (1771–1832) classic description of geologists at work: "Some rin uphill and down dale, knappin' the chucky stones to pieces like sa' many roadmakers run daft. They say it is to see how the warld was made!"

Indeed—to see how the world was made, to see how it continues to evolve, to find its valuable resources, to prevent contamination of its waters and soils, and to predict its dangerous movements. That is why geologists spend months at sea drilling holes in the ocean floor, why they scale mountains, camp in rain-drenched jungles, and trudge through scorching desert winds (**Fig. P.2**). That is why geologists use electron microscopes to examine the atomic structure of minerals, use mass spectrometers to measure the composition of rock and water, and use supercomputers to model the paths of earthquake waves. For over two centuries, geologists have pored over the Earth in search of ideas to explain the processes that form and change our planet.

P.2 The Nature of Geology

Geology, or geoscience, is the study of the Earth. Not only do geologists address academic questions such as the formation and composition of our planet, the causes of earthquakes and ice ages, and the evolution of life, but they also address practical problems such as how to keep pollution out of groundwater, how to find oil and minerals, and how to avoid landslides. And in recent years, geologists have made significant contributions to the study of global climate change. When news reports begin with "Scientists say . . ." and then continue with "an earthquake occurred today off Japan," or "landslides will threaten the city," or "chemicals from the proposed toxic waste dump will ruin the town's water supply," or "there's only a limited supply of oil left," the scientists referred to are geologists.

The fascination of geology attracts many to careers in this science. Tens of thousands of geologists work for oil, mining, water, engineering, and environmental companies, while a smaller number work in universities, government geological surveys, and research laboratories. Nevertheless, since most of the students reading this book will not become professional geologists, it's fair to ask the question, "Why should people, in general, study geology?"

FIGURE P.1 Geologic fieldwork in Antarctica unlocks the mysteries of an icebound continent. A map of Antarctica emphasizes that the Transantarctic Mountains separate West Antarctica from East Antarctica.

A plane dropping geologists on a snowfield

Sledding to a field site with crates of supplies

FIGURE P.2 Geologists in the field. Exploring an outcrop hidden in the woods of Ontario, Canada.

First, geology may be one of the most practical subjects you can learn. Ask yourself the following questions, and you'll realize that geologic processes, phenomena, and materials play major roles in daily life:

> Do you live in a region threatened by landslides, volcanoes, earthquakes, or floods (**Fig. P.3**)?
> Are you worried about the price of energy or about whether there will be a war in an oil-supplying country?
> Do you ever wonder about where the copper in your home's wires comes from?
> Have you seen fields of green crops surrounded by desert and wondered where the irrigation water comes from?
> Would you like to buy a dream house on a beach or near a river?

FIGURE P.3 Human-made cities cannot withstand the vibrations of a large earthquake. These apartment buildings collapsed during an earthquake in Turkey.

> Are you following news stories about how toxic waste can migrate underground into your town's well water?

Clearly, all citizens of the 21st century, not just professional geologists, need to make decisions and understand news reports addressing Earth-related issues. A basic understanding of geology will help you do so.

Second, the study of geology gives you a holistic context for interpreting your surroundings. As you will see, the Earth is a complicated entity, where living organisms, oceans, atmosphere, and solid rock interact with one another in a great variety of ways. Geologic study reveals Earth's antiquity and demonstrates how the planet has changed profoundly during its existence. What our ancestors considered to be the center of the Universe has become, with the development of geologic perspective, our "island in space" today. And what was believed to be an unchanging orb originating at the same time as humanity has become a dynamic planet that existed long before people did and continues to evolve.

Third, the study of geology puts the accomplishments and consequences of human civilization in a broader context. View the aftermath of a large earthquake, flood, or hurricane, and it's clear that the might of natural geologic phenomena greatly exceeds the strength of human-made structures. But watch a bulldozer clear a swath of forest, a dynamite explosion remove the top of a hill, or a prairie field evolve into a housing development, and it's clear that people can change the face of the Earth at rates often exceeding those of natural geologic processes.

Finally, when you finish reading this book, your view of the world may be forever colored by geologic curiosity. If you walk in the mountains, you will think of the many forces that shape and reshape the Earth's surface. If you hear about a natural disaster, you will have insight into the processes that brought it about. And if you go on a road trip, the rock exposures along the highway will no longer be gray, faceless cliffs, but will present complex puzzles of texture and color telling a story of Earth's history.

P.3 Themes of This Book

A number of narrative themes appear (and reappear) throughout this text. These themes, listed below, can be viewed as this book's overall take-home message.

> *The Earth is a unique, evolving system.* Geologists increasingly recognize that the Earth is a complex system; its interior, solid surface, oceans, atmosphere, and life forms interact in many ways to yield the landscapes and environment in which we live. Within this **Earth System**, chemical elements cycle between different types of rock, between rock and sea, between sea and air, and between all of these entities and life.

Heat and Heat Transfer

The atoms and molecules that make up matter are not motionless, but rather jiggle in place and/or move with respect to one another. This activity produces **thermal energy**—the faster the atoms vibrate or move, the greater the thermal energy. Put another way, thermal energy in a substance represents the *sum* of the kinetic energy (energy of motion) of all the substance's atoms.

When we say that one object is hotter or colder than another, we are describing its **temperature**, a measure of warmth relative to some standard. Temperature represents the *average* kinetic energy of atoms in the material. We use the freezing and boiling points of water at sea level as a standard for defining temperatures. In the Celsius (centigrade) scale, the freezing point of water is 0°C and the boiling point is 100°C, whereas in the Fahrenheit scale, the freezing point is 32°F and the boiling point is 212°F. The coldest a substance can

be is the temperature at which its atoms or molecules stand still. We call this temperature absolute zero, or 0K, where K stands for Kelvin, another unit of temperature. Degrees in the Kelvin scale have the same increment as degrees in the Celsius scale; absolute zero equates to −273.15°C. The term **heat** refers to the thermal energy *transferred* from one object to another. Heat can transfer from one place to another by four distinct means:

> Electromagnetic waves transport heat to a body as **radiation**. Radiation traveling from the Sun is responsible for heating the Earth's surface.

> If you stick the end of an iron bar in a fire, **conduction** takes place and heat moves up the bar. This happens because atoms in the bar at the hot end start to vibrate more energetically, and this motion incites atoms farther up the bar to start jiggling. Though heat flows along the bar, the atoms do

not actually move from one locality to another.

> When you place a pot on a stove and heat it, water at the base of the pot gets hotter and expands, so its density decreases. The hot, less-dense water becomes buoyant relative to the colder, denser water, above it. In a gravitational field, buoyant material rises if the material above it is weak enough to flow out of the way. Since liquid water flows easily, the hot water can rise. When this happens, cold water sinks to take its place. The resulting circulation, during which flow of the material itself carries heat, is called **convection**, the path of flow defines **convective cells**.

> **Advection** happens when a hot fluid flows into cracks and pores within a solid, and heats up the solid. The hottness of a metal water pipe's surface, for example, comes from the hot water through the pipe. In the Earth, advection occurs where molten rock rises into the crust.

FIGURE P.4   A simplified map of the Earth's plates. The arrows indicate the direction each plate is moving, and the length of the arrow indicates plate velocity (the longer the arrow, the faster the motion). We discuss the types of plate boundaries in Chapter 2.

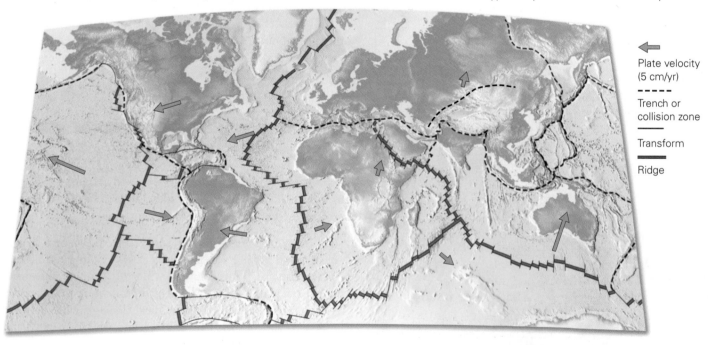

Plate velocity
(5 cm/yr)

Trench or
collision zone

Transform

Ridge

> *Geology helps you understand physical science.* Geology incorporates many of the basic concepts of physics and chemistry because Earth materials are a form of matter, and energy drives geologic processes (**Box P.1**). Thus, studying geology can help you develop a better grasp of key ideas in physical science.

> *Plate tectonics explains many Earth processes.* Earth is not a homogeneous ball, but rather consists of concentric layers—from center to surface, Earth has a core, mantle, and crust. We live on the surface of the crust, where it meets the atmosphere and the oceans. In the 1960s, geologists recognized that the crust, together with the uppermost part of the underlying mantle, forms a 100- to 150-km-thick semi-rigid shell. Large cracks separate this shell into discrete pieces, called **plates**, which move very slowly relative to one another (**Fig. P.4**). The model that describes this movement and its consequences is called the **theory of plate tectonics**, and it is the foundation for understanding most geologic phenomena. Although plates move very slowly—generally less than 10 cm a year—their movements yield earthquakes, volcanoes, and mountain ranges, and cause the map of Earth's surface to change over time.

> *The Earth is a planet.* Despite its uniqueness, the Earth can be viewed as a planet, formed like the other planets of the Solar System from dust and gas that encircled the newborn Sun.

> *The Earth is very old.* Geologic data indicate that the Earth formed 4.57 billion years ago—plenty of time for life to evolve, and for the map of the planet to change. Plate-movement rates of only a few centimeters per year can move a continent thousands of kilometers. There is time enough to build mountains and time enough to grind them down, many times over. To define intervals of this time, geologists developed the **geologic time scale**. **Figure P.5** depicts major subdivisions of the geologic time scale. Chapter 10 discusses these in greater detail.

> *Internal and external processes drive geologic phenomena.* Internal processes are those phenomena driven by heat from inside the Earth. Plate movement is an example. Because plate movements cause mountain building, earthquakes, and volcanoes, we call all of these phenomena internal processes as well. External processes are those phenomena driven by heat supplied by radiation that comes to the Earth from the Sun. This heat drives the movement of air and of water, which grinds and sculpts the Earth's surface and transports the debris to new locations, where it accumulates. The interaction between internal and external processes forms the landscapes of our planet. As we'll see, gravity—the pull that one mass exerts on another—plays an important role in both internal and external processes.

> *Geologic phenomena affect our environment.* Volcanoes, earthquakes, landslides, floods, groundwater, energy sources, and mineral reserves are of vital interest to every inhabitant of this planet. Therefore, throughout this book we emphasize the linkages between geology, the environment, and society.

> *Physical aspects of the Earth System are linked to life processes.* All life on this planet depends on such physical features as the minerals in soil; the temperature, humidity, and composition of the atmosphere; and the flow of surface and subsurface water. And life in turn affects and alters physical features. For example, the oxygen in Earth's atmosphere comes primarily from plant photosynthesis, a life activity.

FIGURE P.5 The geologic time scale.

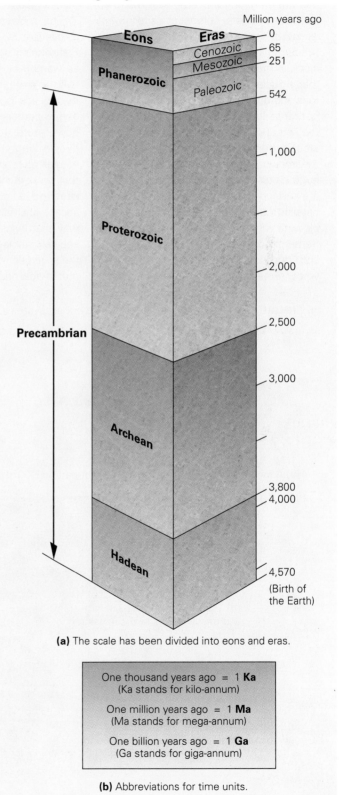

(a) The scale has been divided into eons and eras.

One thousand years ago = 1 **Ka**
(Ka stands for kilo-annum)

One million years ago = 1 **Ma**
(Ma stands for mega-annum)

One billion years ago = 1 **Ga**
(Ga stands for giga-annum)

(b) Abbreviations for time units.

The Scientific Method

Sometime during the past 200 million years, a large block of rock or metal, which had been orbiting the Sun, slammed into our planet. It made contact at a site in what is now the central United States, a landscape of flat cornfields. The impact of this block, a meteorite, released more energy than a nuclear bomb—a cloud of shattered rock and dust blasted skyward, and once-horizontal layers of rock from deep below the ground sprang upward and tilted steeply beneath the gaping hole left by the impact. When the dust had settled, a huge crater surrounded by debris marked the surface of the Earth at the impact site. Later in Earth history, running water and blowing wind wore down this jagged scar. Some 15,000 years ago, sand, gravel, and mud carried by a vast glacier buried what remained, hiding it entirely from

view (**Fig. BxP.1a–c**). Wow! So much history beneath a cornfield. How do we know this? It takes scientific investigation.

The movies often portray science as a dangerous tool, capable of creating Frankenstein's monster, and scientists as warped or nerdy characters with thick glasses and poor taste in clothes. In reality, science is simply the use of observation, experiment, and calculation to explain how nature operates, and scientists are people who study and try to understand natural phenomena. Scientists carry out their work in the context of the **scientific method**, a sequence of steps for systematically analyzing scientific problems in a way that leads to verifiable results. Let's see how geologists employed the steps of the scientific method to come up with the meteorite-impact story.

1. *Recognizing the problem.* Any scientific project, like any detective story, begins by identifying a mystery. The cornfield mystery came to light when water drillers discovered limestone, a rock typically made of shell fragments, just below the 15,000-year-old glacial sediment. In surrounding regions, the rock beneath the glacial sediment consists of sandstone, a rock made of cemented-together sand grains. Since limestone can be used to build roads, make cement, and produce the agricultural lime used in treating soil, workers stripped off the glacial sediment and dug a quarry to excavate the limestone. They were amazed to find that rock layers exposed in the quarry were tilted steeply and had

FIGURE BxP.1 An ancient meteorite impact excavates a crater and permanently changes rock beneath the surface.

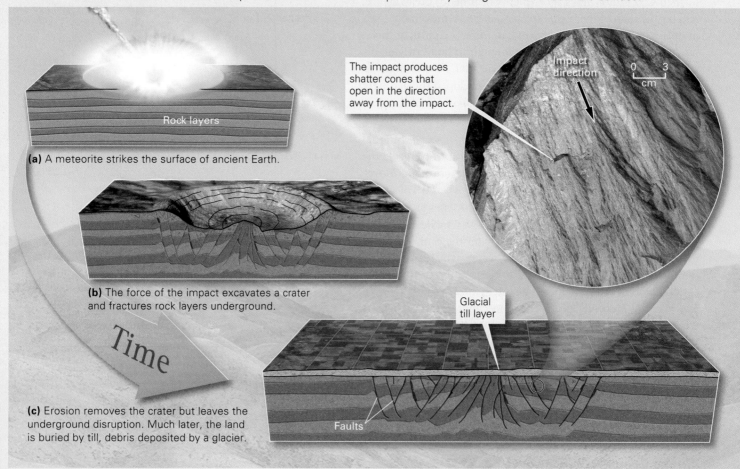

The impact produces shatter cones that open in the direction away from the impact.

Impact direction

Rock layers

(a) A meteorite strikes the surface of ancient Earth.

(b) The force of the impact excavates a crater and fractures rock layers underground.

Time

Glacial till layer

Faults

(c) Erosion removes the crater but leaves the underground disruption. Much later, the land is buried by till, debris deposited by a glacier.

been shattered by large cracks. In the surrounding regions, all rock layers are horizontal like the layers in a birthday cake, and the rocks contain relatively few cracks. Curious geologists came to investigate and soon realized that the geologic features of the land just beneath the cornfield presented a problem to be explained: What phenomena had brought limestone up close to the Earth's surface, had tilted the layering in the rocks, and had shattered the rocks?

2. *Collecting data.* The scientific method proceeds with the collection of observations or clues that point to an answer. Geologists studied the quarry and determined the age of its rocks, measured the orientation of the rock layers, and documented (made a written or photographic record of) the fractures that broke up the rocks.

3. *Proposing hypotheses.* A scientific **hypothesis** is merely a possible explanation, involving only natural processes, that can explain a set of observations. Scientists propose hypotheses during or after their initial

data collection. In this example, the geologists working in the quarry came up with two alternative hypotheses. First, the features in this region could result from a volcanic explosion; and second, they could result from a meteorite impact.

4. *Testing hypotheses.* Since a hypothesis is no more than an idea that can be either right or wrong, scientists must put hypotheses through a series of tests to see if they work. The geologists at the quarry compared their field observations with published observations made at other sites of volcanic explosions and meteorite impacts, and they studied the results of experiments designed to simulate such events. If the geologic features visible in the quarry were the result of volcanism, the quarry should contain rocks formed by solidification of molten rock erupted by a volcano. But no such rocks were found. If, however, the features were the result of an impact, the rocks should contain **shatter cones**, small, cone-shaped cracks (see Fig. Bx P.1c).

Shatter cones can be overlooked, so the geologists returned to the quarry specifically to search for them, and found them in abundance. The impact hypothesis passed the test!

Theories are scientific ideas supported by abundant evidence; they have passed many tests and have failed none. Scientists are much more confident in the correctness of a theory than of a hypothesis. Continued study in the quarry eventually yielded so much evidence for impact that the impact hypothesis came to be viewed as a theory. Scientists continue to test theories over a long time. Successful theories withstand these tests and are supported by so many observations that they become part of a discipline's foundation. However, some theories may be disproven and replaced by better ones.

In a few cases, scientists have been able to devise concise statements that completely describe a specific relationship or phenomenon. Such statements are called **scientific laws**. Note that the *law* of gravity does not explain why gravity exists, but the *theory* of evolution does provide an explanation of why evolution occurs.

> *Science comes from observation, and people make scientific discoveries.* Science does not consist of subjective guesses or arbitrary dogmas, but rather of a consistent set of objective statements resulting from the application of the **scientific method** (**Box P.2**). Every scientific idea must be tested thoroughly, and should be used only when supported by documented observations. Further, scientific ideas do not appear out of nowhere, but are the result of human efforts. Wherever possible, this book shows where geologic ideas came from, and tries to answer the question, "How do we know that?"

As you read this book, please keep these themes in mind. Don't view geology as a list of words to memorize, but rather as an interconnected set of concepts to digest. Most of all, enjoy yourself as you learn about what may be the most fascinating planet in the Universe. To help illustrate the geology of our amazing world, we have created "See for Yourself" features. Using *Google Earth*™, you'll be able to find examples of localities that illustrate geologic features and phenomena.

Key Terms

advection (p. 4)	geologic time scale (p. 5)	plate (p. 5)	shatter cones (p. 7)
conduction (p. 4)	geologist (p. 2)	radiation (p. 4)	temperature (p. 4)
convection (p. 4)	geology (p. 2)	scientific laws (p. 7)	theory (p. 7)
convection cells (p. 4)	heat (p. 4)	scientific method	theory of plate tectonics (p. 5)
Earth System (p. 3)	hypothesis (p. 7)	(p. 6)	thermal energy (p. 4)

The Earth in Context

Chapter Objectives

By the end of this chapter you should know . . .

> modern concepts concerning the basic architecture of our Universe and its components.

> the character of our own Solar System.

> scientific explanations for the formation of the Universe and the Earth.

> the overall character of the Earth's magnetic field, atmosphere, and surface.

> the variety and composition of materials that make up our planet.

> the nature of Earth's internal layering.

This truth within thy mind rehearse,
That in a boundless Universe
Is boundless better, boundless worse.

—Alfred, Lord Tennyson (British poet, 1809–1892)

1.1 Introduction

Sometime in the distant past, humans developed the capacity for complex, conscious thought. This amazing ability, which distinguishes our species from all others, brought with it the gift of curiosity, an innate desire to understand and explain the workings of ourselves and of all that surrounds us—our Universe. Astronomers define the **Universe** as all of space and all the matter and energy within it. Questions that we ask about the Universe differ little from questions a child asks of a playmate: Where do you come from? How old are you? Such musings first spawned legends in which heroes, gods, and goddesses used supernatural powers to mold the planets and sculpt the landscape. Eventually, researchers began to apply scientific principles to **cosmology**, the study of the overall structure and history of the Universe.

In this chapter, we begin with a brief introduction to the principles of scientific cosmology—we characterize the basic architecture of the Universe, introduce the Big Bang theory for the formation of the Universe, and discuss scientific ideas concerning the birth of the Earth. Then we outline the basic characteristics of our home planet by building an image of its surroundings, surface, and interior. Our high-speed tour of the Earth provides a reference frame for the remainder of this book.

1.2 An Image of Our Universe

What Is the Structure of the Universe?

Think about the mysterious spectacle of a clear night sky. What objects are up there? How big are they? How far away are they? How do they move? How are they arranged? In addressing such questions, ancient philosophers first distinguished between stars (points of light whose locations relative to each other are fixed) and planets (tiny spots of light that move relative to the backdrop of stars). Over the centuries, two schools of thought developed concerning how to explain the configuration of stars and planets, and their relationships to the Earth, Sun, and Moon. The first school advocated a **geocentric model (Fig. 1.1a)**, in which the Earth sat without moving at the center of the Universe, while the Moon and the planets whirled around it within a revolving globe of stars. The second school advocated a **heliocentric model (Fig. 1.1b)**, in which the Sun lay at the center of the Universe, with the Earth and other planets orbiting around it.

The geocentric image eventually gained the most followers, due to the influence of an Egyptian mathematician, Ptolemy (100–170 C.E.), for he developed equations that appeared to predict the wanderings of the planets in the context of the geocentric model. During the Middle Ages (ca. 476–1400 C.E.), church leaders in Europe adopted Ptolemy's geocentric model as dogma, because it justified the comforting thought that humanity's home occupies the most important place in the Universe. Anyone who disagreed with this view risked charges of heresy.

Then came the Renaissance. In 15th-century Europe, bold thinkers spawned a new age of exploration and scientific discovery. Thanks to the efforts of Nicolaus Copernicus (1473–1543) and Galileo Galilei (1564–1642), people gradually came to realize that the Earth and planets did indeed orbit the Sun and could not be at the center of the Universe. And when Isaac Newton (1643–1727) explained **gravity**, the attractive force that one object exerts on another, it finally became possible to understand why these objects follow the orbits that they do.

In the centuries following Newton, scientists gradually adopted modern terminology for discussing the Universe. In this language, the Universe contains two related entities: matter and energy. *Matter* is the substance of the Universe—it takes up space and you can feel it. We refer to the amount of matter in an object as its *mass*, so an object with greater mass contains more matter. *Density* refers to the amount of mass occupying a given volume of space. The mass of an object determines its *weight*, the force that acts on an object due to gravity.

An object always has the same mass, but its weight varies depending on where it is. For example, on the Moon, you weigh much less than on the Earth. The matter in the Universe does not sit still. Components vibrate and spin, they move from one place to another, they pull on or push against each other, and they break apart or combine. In a general sense, we consider such changes to be kinds of "work." Physicists refer

FIGURE 1.1 Contrasting views of the Universe, as drawn by artists hundreds of years ago.

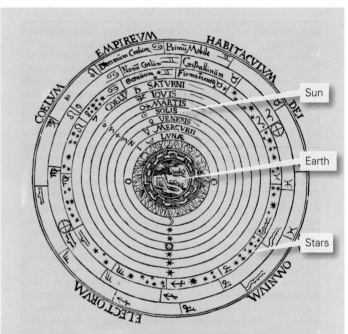

(a) The *geocentric* image of the Universe shows the Earth at the center, surrounded by air, fire, and the other planets, all contained within the globe of the stars.

(b) The *heliocentric* image of the Universe shows the Sun at the center, as envisioned by Copernicus.

FIGURE 1.2 A galaxy may contain about 300 billion stars.

The Milky Way, as viewed from Earth.

A galaxy that looks like the Milky Way.

(a) A Hubble image reveals many galaxies in what looks like empty space to our naked eyes.

(b) As seen from the Earth, the Milky Way looks like a hazy cloud.

(c) From space, it would look like a giant spiral.

to the ability to do work as **energy**. One piece of matter can do work directly on another by striking it. Heat, light, magnetism, and gravity all provide energy that can cause change at a distance.

As the understanding of matter and energy improved, and telescopes became refined so that astronomers could see and measure features progressively farther into space, the interpretation of stars evolved. Though it looks like a point of light, a **star** is actually an immense ball of incandescent gas that emits intense heat and light. Stars are not randomly scattered through the Universe; gravity holds them together in immense groups called **galaxies**. The Sun and over 300 billion stars together form the Milky Way galaxy. More than 100 billion galaxies constitute the visible Universe (**Fig. 1.2a**).

From our vantage point on Earth, the Milky Way looks like a hazy band (**Fig. 1.2b**), but if we could view the Milky Way from a great distance, it would look like a flattened spiral with great curving arms slowly swirling around a glowing, disk-like center (**Fig. 1.2c**). Presently, our Sun lies near the outer edge of one of these arms and rotates around the center of the galaxy about once every 250 million years. So, we hurtle through space, relative to an observer standing outside the galaxy, at about 200 km per second.

Clearly, human understanding of Earth's place in the Universe has evolved radically over the past few centuries. Neither

Did you ever wonder...
how fast you are traveling through space?

the Earth, nor the Sun, nor even the Milky Way occupy the center of the Universe—and everything is in motion.

The Nature of Our Solar System

Eventually, astronomical study demonstrated that our Sun is a rather ordinary, medium-sized star. It looks like a sphere, instead of a point of light, because it is much closer to the Earth than are the stars. The Sun is "only" 150 million km (93 million miles) from the Earth. Stars are so far away that we measure their distance in light years, where 1 light year is the distance traveled by light in one year, about 10 trillion km, or 6 trillion miles—the nearest star beyond the Sun is over 4 light years away. How can we picture distances? If we imagined that the Sun were the size of a golf ball (about 4.3 cm), then the Earth would be a grain of sand about one meter away, and the nearest star would be 270 km (168 miles) away. (Note that the distance between stars is tiny by galactic standards—the Milky Way galaxy is 120,000 light years across!)

Our Sun is not alone as it journeys through the heavens. Its gravitational pull holds on to many other objects which, together with the Sun, comprise the **Solar System** (**Fig. 1.3a, b**). The Sun accounts for 99.8% of the mass in the Solar System. The remaining 0.2% includes a great variety of objects, the largest of which are planets. Astronomers define a **planet** as an object that orbits a star, is roughly spherical, and has "cleared its neighborhood of other objects." The last phrase in this definition sounds a bit strange at first, but merely implies that a planet's gravity has pulled in all particles of matter in its orbit.

FIGURE 1.3 The relative sizes and positions of planets in the Solar System.

(a) Relative sizes of the planets. All are much smaller than the Sun, but the gas-giant planets are much larger than the terrestrial planets. Jupiter's diameter is about 11.2 times greater than that of Earth.

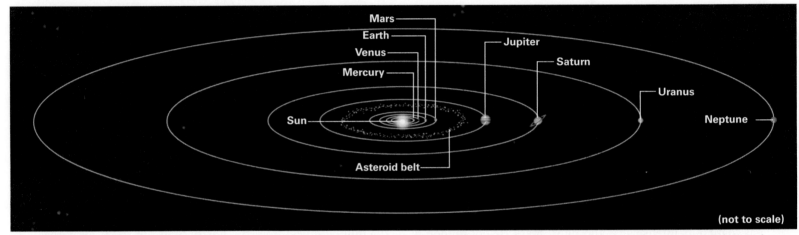

(b) Relative positions of the planets. This figure is not to scale. If the Sun in this figure was the size of a large orange, the Earth would be the size of a sesame seed 15 meters (49 feet) away. Note that all planetary orbits lie roughly in the same plane.

According to this definition, which was formalized in 2005, our Solar System includes eight planets—Mercury, Venus, Earth, Mars, Jupiter, Saturn, Uranus, and Neptune. In 1930, astronomers discovered Pluto, a 2,390-km-diameter sphere of ice, whose orbit generally lies outside that of Neptune's. Until 2005, astronomers considered Pluto to be a planet. But since it does not fit the modern definition, it has been dropped from the roster. Our Solar System is not alone in hosting planets; in recent years, astronomers have found planets orbiting stars in many other systems. As of 2012, over 760 of these "exoplanets" have been found.

Planets in our Solar System differ radically from one another both in size and composition. The inner planets (Mercury, Venus, Earth, and Mars), the ones closer to the Sun, are relatively small. Astronomers commonly refer to these as **terrestrial planets** because, like Earth, they consist of a shell of rock surrounding a ball of metallic iron alloy. The outer planets (Jupiter, Saturn, Uranus, and Neptune) are known as the **giant planets**, or Jovian planets. The adjective *giant* certainly seems appropriate, for these planets are huge—Jupiter, for example, has a mass 318 times larger than that of Earth and accounts for about 71% of the non-solar mass in the Solar System. The overall composition of the giant planets is very different from that of the terrestrial planets. Specifically, most of the mass of Neptune and Uranus contain solid forms of water, ammonia, and methane, so these planets are known as the *ice giants*. Most of the mass of Jupiter and Saturn consists of hydrogen and helium gas or liquefied gas, so these planets are known as the *gas giants*.

In addition to the planets, the Solar System contains a great many smaller objects. Of these, the largest are moons. A **moon** is a sizable body locked in orbit around a planet. All but two planets (Mercury and Venus) have moons in varying numbers— Earth has one, Mars has two, and Jupiter has at least 63. Some moons, such as Earth's Moon, are large and spherical, but most are small and have irregular shapes. In addition to moons, millions of asteroids (chunks of rock and/or metal) comprise a belt between the orbits of Mars and Jupiter. Asteroids range in size from less than a centimeter to about 930 km in diameter. And about a trillion bodies of ice lie in belts or clouds beyond the orbit of Neptune. Most of these icy objects are tiny, but a few (including Pluto) have diameters of over 2,000 km and may be thought of as "dwarf planets." The gravitational pull of the main planets has sent some of the icy objects on paths that take them into the inner part of the Solar System, where they begin to evaporate and form long tails of gas—we call such objects comets.

> ## Take-Home Message
> The Earth is one of eight planets (four terrestrial and four gas or ice giants) orbiting our Sun, which is one of 300 billion stars of the revolving, spiral-shaped Milky Way galaxy. Hundreds of billions of galaxies speckle the visible Universe.

1.3 Forming the Universe

We stand on a planet, in orbit around a star, speeding through space on the arm of a galaxy. Beyond our galaxy lie hundreds of billions of other galaxies. Where did all this "stuff"—the matter of the Universe—come from, and when did it first form? For most of human history, a scientific solution to these questions seemed intractable. But in the 1920s, unexpected observations about the nature of light from distant galaxies set astronomers on a path of discovery that ultimately led to a model of Universe formation known as the Big Bang theory. To explain these observations, we must first introduce an important phenomenon called the Doppler effect. We then show how this understanding leads to the recognition that the Universe is expanding, and finally, to the conclusion that this expansion began during the Big Bang, 13.7 billion years ago.

Waves and the Doppler Effect

When a train whistle screams, the sound you hear moved through the air from the whistle to your ear in the form of sound waves. **Waves** are disturbances that transmit energy from one point to another in the form of periodic motions. As each sound wave passes, air alternately compresses, then expands. We refer to the distance between successive waves as the **wavelength**, and the number of waves that pass a point in a given time interval as the **frequency**. If the wavelength decreases, more waves pass a point in a given time interval, so the frequency increases. The pitch of a sound, meaning its note on the musical scale, depends on the frequency of the sound waves.

Imagine that you are standing on a station platform while a train moves toward you. The train whistle's sound gets louder as the train approaches, but its pitch remains the same. The instant the train passes, the pitch abruptly changes—it sounds like a lower note in the musical scale. Why? When the train moves toward you, the sound has a higher frequency (the waves are closer together so the wavelength is smaller) because the sound source, the whistle, has moved slightly closer to you between the instant that it emits one wave and the instant that it emits the next (**Fig. 1.4a, b**). When the train moves away from you, the sound has a lower frequency (the waves are farther apart), because the whistle has moved slightly farther from you between the instant it emits one wave and the instant it emits the next. An Austrian physicist, C. J. Doppler (1803–1853), first interpreted this phenomenon, and thus the change in frequency that happens when a wave source moves is now known as the **Doppler effect**.

Light energy also moves in the form of waves. We can represent light waves symbolically by a periodic succession of crests and troughs (**Fig. 1.4c**). Visible light comes in many colors—the colors of the rainbow. The color you see depends on the frequency of the light waves, just as the pitch of a sound you hear depends on the frequency of sound waves. Red light has a longer wavelength (lower frequency) than does blue light. The Doppler effect also applies to light but can be noticed only if the light source moves very fast, at least a few percent of the speed of light. If a light source moves away from you, the light you see becomes redder, as the light shifts to longer wavelength or lower frequency. If the source moves toward you, the light you see becomes bluer, as the light shifts to higher frequency. We call these changes the **red shift** and the blue shift, respectively.

Does the Size of the Universe Change?

In the 1920s, astronomers such as Edwin Hubble, after whom the Hubble Space Telescope was named, braved many a frosty night beneath the open dome of a mountaintop observatory in order to aim telescopes into deep space. These researchers were searching for distant galaxies. At first, they documented

FIGURE 1.4 Manifestations of the Doppler effect for sound and for light.

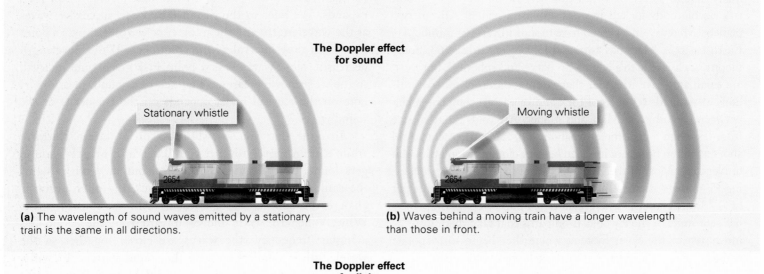

The Doppler effect for sound

Stationary whistle

Moving whistle

(a) The wavelength of sound waves emitted by a stationary train is the same in all directions.

(b) Waves behind a moving train have a longer wavelength than those in front.

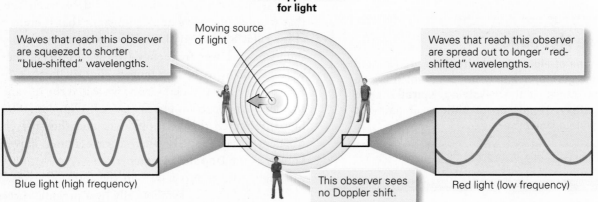

The Doppler effect for light

Moving source of light

Waves that reach this observer are squeezed to shorter "blue-shifted" wavelengths.

Waves that reach this observer are spread out to longer "red-shifted" wavelengths.

Blue light (high frequency)

This observer sees no Doppler shift.

Red light (low frequency)

(c) The wavelength of blue light is less than that of red light. If a light source moves very fast, the Doppler effect results in a shifting of the wavelengths. The observed shift depends on the position of the observer.

Sun

Distant galaxy

(d) The atoms in a star absorb certain specific wavelengths of light. We see these wavelengths as dark lines on a light spectrum. Note that the lines from a galaxy a billion light years away are shifted toward the red end of the spectrum (i.e., to the right), in comparison to the lines from our own Sun.

only the location and shape of newly discovered galaxies, but eventually they also began to study the wavelength of light produced by the distant galaxies. The results yielded a surprise that would forever change humanity's perception of the Universe. To their amazement, the astronomers found that the light of distant galaxies *display a red shift* relative to the light of a nearby star (**Fig. 1.4d**).

Hubble pondered this mystery and, around 1929, attributed the red shift to the Doppler effect, and concluded that the distant galaxies must be moving away from Earth at an immense velocity. At the time, astronomers thought the Universe had a fixed size, so Hubble initially assumed that if some galaxies were moving away from

Earth, others must be moving toward Earth. But this was not the case. On further examination, Hubble concluded that the light from all distant galaxies, regardless of their direction from Earth, exhibits a red shift. In other words, *all* distant galaxies are moving rapidly away from us.

How can all galaxies be moving away from us, regardless of which direction we look? Hubble puzzled over this question and finally recognized the solution: the whole Universe must be expanding! To picture the expanding Universe, imagine a ball of bread dough with raisins scattered throughout. As the

Did you ever wonder...
do galaxies move?

FIGURE 1.5 The concept of the expanding Universe and the Big Bang.

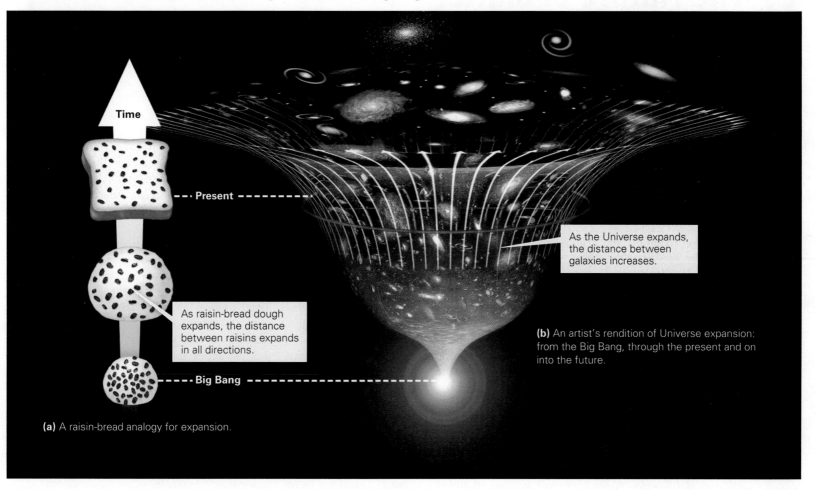

Time

Present

As raisin-bread dough expands, the distance between raisins expands in all directions.

Big Bang

(a) A raisin-bread analogy for expansion.

As the Universe expands, the distance between galaxies increases.

(b) An artist's rendition of Universe expansion: from the Big Bang, through the present and on into the future.

dough bakes and expands into a loaf, each raisin moves away from its neighbors, in every direction; **Fig. 1.5a.** This idea came to be known as the **expanding Universe theory**.

The Big Bang

Hubble's ideas marked a revolution in cosmological thinking. Now we picture the Universe as an expanding bubble, in which galaxies race away from each other at incredible speeds. This image immediately triggers the key question of cosmology: did the expansion begin at some specific time in the past? If it did, then that instant would mark the physical beginning of the Universe.

Most astronomers have concluded that expansion did indeed begin at a specific time, with a cataclysmic explosion called the Big Bang. According to the **Big Bang theory**, all matter and energy—everything that now constitutes the Universe—was initially packed into an infinitesimally small point. The point "exploded" and the Universe began, according to current estimates, 13.7 (±1%) billion years ago.

Of course, no one was present at the instant of the Big Bang, so no one actually saw it happen. But by combining clever calculations with careful observations, researchers have developed a consistent model of how the Universe evolved, beginning an instant after the explosion (**Fig. 1.5b**). According to this model of the Big Bang, profound change happened at a fast and furious rate at the outset. During the first instant of existence, the Universe was so small, so dense, and so hot that it consisted entirely of energy—atoms, or even the smallest subatomic particles that make up atoms, could not even exist. (See **Box 1.1** for a review of atomic structure.) Within a few seconds, however, hydrogen atoms could begin to form. And by the time the Universe reached an age of 3 minutes, when its temperature had fallen below 1 billion degrees, and its diameter had grown to about 53 million km (35 million miles), hydrogen atoms could fuse together to form helium atoms. Formation of new nuclei in the first few minutes of time is called *Big Bang nucleosynthesis* because it happened *before* any stars existed. This process could produce only light atoms, meaning ones containing a small number of protons (an atomic number less than 5), and

The Nature of Matter

What does matter consist of? A Greek philosopher named Democritus (ca. 460–370 B.C.E.) argued that if you kept dividing matter into progressively smaller pieces, you would eventually end up with nothing; but since it's not possible to make something out of nothing, there must be a smallest piece of matter that can't be subdivided further. He proposed the name "atom" for these smallest pieces, based on the Greek word *atomos*, which means indivisible.

Our modern understanding of matter developed in the 17th century, when chemists recognized that certain substances (such as hydrogen and oxygen) *cannot* break down into other substances, whereas others (such as water and salt) *can* break down. The former came to be known as *elements*, and the latter came to be known as *compounds*. John Dalton (1766–1844) adopted the word *atom* for the smallest piece of an element that has the property of the element; the smallest piece of a compound that has the properties of the compound is a *molecule*. Separate atoms are held together to form molecules by *chemical bonds*, which we discuss more fully later in the book. As an example, chemical bonds hold two hydrogen atoms to form an H_2 molecule.

Chemists in the 17th and 18th centuries identified 92 naturally occurring elements on Earth; modern physicists have created more than a dozen new ones. Each element has a name and a symbol (e.g., N = nitrogen; H = hydrogen; Fe = iron; Ag = silver).

Atoms are so small that over five trillion (5,000,000,000,000) can fit on the head of a pin. Nevertheless, in 1910, Ernest Rutherford, a British physicist, proved that, contrary to the view of Democritus, atoms actually can be divided into smaller pieces. Most of the mass in an atom clusters in a dense ball, called the *nucleus*, at the atom's center. The nucleus contains two types of subatomic particles: *neutrons*, which have a neutral electrical charge, and *protons*, which have a positive charge. A cloud of electrons surrounds the nucleus (**Fig. Bx1.1a**); an *electron* has a negative charge and contains only 1/1,836 as much mass as a proton. ("Charge," simplistically, refers to the way in which a particle responds to a magnet or an electric current.) Roughly speaking, the diameter of an electron cloud is 10,000 times greater than that of the nucleus, yet the cloud contains only 0.05% of an atom's mass—thus, atoms are mostly empty space!

We distinguish atoms of different elements from one another by their *atomic number*, the number of protons in their nucleus. Smaller atoms have smaller atomic numbers, and larger ones have larger atomic numbers. The lightest atom, hydrogen, has an atomic number of 1, and the heaviest naturally occurring atom, uranium, has an atomic number of 92. Except for hydrogen nuclei, all nuclei also contain neutrons. In smaller atoms, the number of neutrons roughly equals the number of protons, but in larger atoms the number of neutrons exceeds the number of protons. The *atomic mass* of an atom is roughly the sum of the number of neutrons and the number of protons. For example, an oxygen nucleus contains 8 protons and 8 neutrons, and thus has an atomic mass of 16.

In 1869, a Russian chemist named Dmitri Mendelév (1834–1907) recognized that groups of elements share similar characteristics, and he organized the elements into a chart that we now call the periodic table of the elements. With modern understanding of the periodic table, it became clear that the ordering of the elements reflects their atomic number and the stream of the electron cloud.

Nuclear bonds serve as the "glue" that holds together subatomic particles in a nucleus. Atoms can change only during nuclear reactions, when nuclear bonds break or form. Physicists recognize several types of nuclear reactions. For example, during "radioactive decay" reactions, a nucleus either emits a subatomic particle or undergoes fission. As a result of **fission**, a large nucleus breaks apart to form two smaller atoms (**Fig. Bx1.1b**). Radioactive decay transforms an atom of one element into an atom of another and produces energy. For example, fission reactions provide the energy of atomic bombs and nuclear power plants. Atoms that spontaneously undergo the process are known as **radioactive elements**.

During **fusion**, smaller atoms collide and stick together to form a larger atom. For example, successive fusion reactions produce a helium atom out of four hydrogen atoms. Fusion reactions power the Sun and occur during the explosion of a hydrogen bomb (**Fig. Bx1.1c**).

FIGURE Bx1.1 The nature of atoms and nuclear reactions.

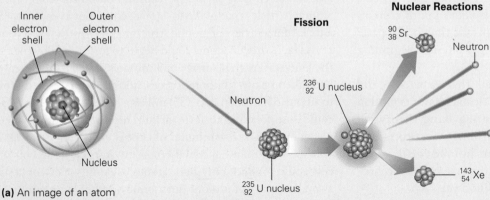

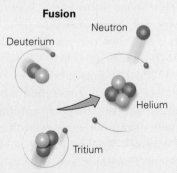

(a) An image of an atom with a nucleus orbited by electrons.

(b) A uranium atom splits during nuclear fission.

(c) Two atoms (versions of hydrogen) stick together to form one atom of helium during nuclear fusion in a hydrogen bomb.

it happened very rapidly. In fact, virtually all of the new atomic nuclei that would form by Big Bang nucleosynthesis existed by the end of the first 5 minutes.

Eventually, the Universe became cool enough for chemical bonds to bind atoms of certain elements together in molecules. Most notably, two hydrogen atoms could join to form molecules of H_2. As the Universe continued to expand and cool further, atoms and molecules slowed down and accumulated into patchy clouds called **nebulae**. The earliest nebulae of the Universe consisted almost entirely of hydrogen (74%, by volume) and helium (24%) gas.

Birth of the First Stars

When the Universe reached its 200 millionth birthday, it contained immense, slowly swirling, dark nebulae separated by vast voids of empty space. The Universe could not remain this way forever, though, because of the invisible but persistent pull of gravity. Eventually, gravity began to remold the Universe pervasively and permanently.

All matter exerts gravitational pull—a type of force—on its surroundings, and as Isaac Newton first pointed out, the amount of pull depends on the amount of mass; the larger the mass, the greater its pull. Somewhere in the young Universe, the gravitational pull of an initially more massive region of a nebula began to suck in surrounding gas and, in a grand example of the rich getting richer, grew in mass and, therefore, density. As this denser region attracted progressively more gas, the gas compacted into a smaller region, and the initial swirling movement of gas transformed into a rotation around an axis. As gas continued to move inward, cramming into a progressively smaller volume, the rotation rate became faster and faster. (A similar phenomenon happens when a spinning ice skater pulls her arms inward.) Because of its increased rotation, the nebula evolved into a disk shape (see **Geology at a Glance**, pp. 22–23). As more and more matter rained down onto the disk, it continued to grow, until eventually, gravity collapsed the inner portion of the disk into a dense ball. As the gas squeezed into a smaller and smaller space, its temperature increased dramatically. Eventually, the central ball of the disk became hot enough to glow, and at this point it became a **protostar**. The remaining mass of the disk, as we will see, eventually clumped into smaller spheres, the planets.

A protostar continues to grow, by pulling in successively more mass, until its core becomes extremely dense and its temperature reaches about 10 million degrees. Under such conditions, hydrogen nuclei slam together so forcefully that they join or "fuse," in a series of steps, to form helium nuclei (see Box 1.1). Such fusion reactions produce huge amounts of energy, and the mass becomes a fearsome furnace. When the first nuclear fusion reactions began in the first protostar, the body "ignited" and the first true star formed. When this happened, perhaps

800 million years after the Big Bang, the first starlight pierced the newborn Universe. This process would soon happen again and again, and many first-generation stars came into existence (see Chapter 1 opener photo).

First-generation stars tended to be very massive, perhaps 100 times the mass of the Sun. Astronomers have shown that the larger the star, the hotter it burns and the faster it runs out of fuel and dies. A huge star may survive only a few million years to a few tens of millions of years before it becomes a **supernova**, a giant explosion that blasts much of the star's matter back into space. Thus, not long after the first generation of stars formed, the Universe began to be peppered with the first generation of supernovas.

> ### Take-Home Message
> According to the Big Bang theory, a cataclysmic explosion at 13.7 Ga formed the Universe, which has been expanding ever since. Atoms formed during the Big Bang collected into nebulae which, due to gravity, collapsed into dense balls, the first stars.

1.4 We Are All Made of Stardust

Where Do Elements Come From?

Nebulae from which the first-generation stars formed consisted entirely of the lightest atoms, because only these atoms were generated by Big Bang nucleosynthesis. In contrast, the Universe of today contains 92 naturally occurring elements. Where did the other 87 elements come from? In other words, how did elements with larger atomic numbers (such as carbon, sulfur, silicon, iron, gold, and uranium), which are common on Earth, form? Physicists have shown that these elements form during the life cycle of stars, by the process of **stellar nucleosynthesis**.

> **Did you ever wonder...**
> where the atoms in your body first formed?

Because of stellar nucleosynthesis, we can consider stars to be "element factories," constantly fashioning larger atoms out of smaller atoms.

What happens to the atoms formed in stars? Some escape into space during the star's lifetime, simply by moving fast enough to overcome the star's gravitational pull. The stream of atoms emitted from a star during its lifetime is a **stellar wind** (**Fig. 1.6a**). Some escape only when a star dies. A small or medium star (like our Sun) releases a large shell of gas as it dies, ballooning into a "red giant" during the process, whereas a large star blasts matter into space during a supernova explosion (**Fig. 1.6b**). Most very heavy atoms (those with atomic

FIGURE 1.6 Element factories in space.

(a) The solar wind ejects matter from the Sun into space.

(b) This expanding cloud of gas, ejected into space from an explosion whose light reached the Earth in 1054 c.e., is called the Crab Nebula.

numbers greater than that of iron) require even more violent circumstances to form than generally occurs within a star. In fact, most very heavy atoms form *during* a supernova explosion. Once ejected into space, atoms from stars and supernova explosions form new nebulae or mix back into existing nebulae.

When the first generation of stars died, they left a legacy of new, heavier elements that mixed with residual gas from the Big Bang. A second generation of stars and associated planets formed out of these compositionally more diverse nebulae. Second-generation stars lived and died, and contributed heavier elements to third-generation stars. Succeeding generations contain a greater proportion of heavier elements. Because not all stars live for the same duration of time, at any given moment the Universe contains many different generations of stars. Our Sun may be a third-, fourth-, or fifth-generation star. Thus, the mix of elements we find on Earth includes relicts of primordial gas from the Big Bang as well as the disgorged guts of dead stars. Think of it—the elements that make up your body once resided inside a star!

The Nebular Theory for Forming the Solar System

Earlier in this chapter, we introduced scientific concepts of how stars form from nebulae. But we delayed our discussion of how the planets and other objects in our Solar System originated until we had discussed the production of heavier atoms such as carbon, silicon, iron, and uranium, because planets consist predominantly of these elements. Now that we've discussed stars as element factories, we return to the early history of the Solar System and introduce the **nebular theory,** an explanation for the origin of planets, moons, asteroids, and comets. According to the nebular theory, these objects formed from the material in the flattened outer part of the disk, the material that did not become part of the star. This outer part is called the **protoplanetary disk**.

What did the protoplanetary disk consist of? The disk from which our Solar System formed contained all 92 elements, some as isolated atoms, and some bonded to others in molecules. Geologists divide the material formed from these atoms and molecules into two classes. **Volatile** materials—such as hydrogen, helium, methane, ammonia, water, and carbon dioxide—are materials that can exist as gas at the Earth's surface. In the pressure and temperature conditions of space, all volatile materials remain in a gaseous state closer to the Sun. But beyond a distance called the "frost line," some volatiles condense into ice. (Note that we do not limit use of the word "ice" to water alone.) **Refractory** materials are those that melt only at high temperatures, and they condense to form solid soot-sized particles of "dust" in the coldness of space. As the proto-Sun began to form, the inner part of the disk became hotter, causing volatile elements to evaporate and drift to the outer portions of the disk. Thus, the inner part of the disk ended up consisting predominantly of refractory dust, whereas the outer portions accumulated large quantities of volatile materials and ice. As this was happening, the protoplanetary disk evolved into a series of concentric rings in response to gravity.

How did the dusty, icy, and gassy rings transform into planets? Even before the proto-Sun ignited, the material of the surrounding rings began to clump and bind together, due to gravity and electrical attraction. First, soot-sized particles merged to form sand- to marble-sized grains that resembled "dust bunnies." Then, these grains stuck together to form grainy basketball-sized blocks (**Fig. 1.7**), which in turn collided. If the collision was slow, blocks stuck together or simply bounced apart. If the collision was fast, one or both of the blocks shattered, producing smaller fragments that recombined later. Eventually, enough blocks coalesced to form **planetesimals**, bodies whose diameter exceeded about 1 km. Because of their mass, the planetesimals exerted enough gravity to attract and pull in other objects that were nearby (see Geology at a Glance, pp. 22–23). Figuratively, planetesimals acted like vacuum cleaners, sucking in small pieces of dust and ice as well as smaller planetesimals that lay in their orbit, and in the process they grew progressively larger. Eventually, victors in the competition to attract mass grew into **protoplanets**, bodies approaching the size of today's planets. Once a protoplanet succeeded in incorporating virtually all the debris within its orbit, it became a full-fledged planet.

Early stages in Earth's planet-forming process probably occurred very quickly—some computer models suggest that it may have taken less than a million years to go from the dust and gas stage to the large planetesimal stage. Planets may have grown from planetesimals in 10 to 200 million years. In the inner orbits, where the protoplanetary disk consisted mostly of dust, small terrestrial planets composed of rock and metal formed. In the outer part of the Solar System, where significant amounts of ice existed, protoplanets latched on to vast amounts of ice and gas and evolved into the giant planets. Fragments of materials that were not incorporated in planets remain today as asteroids and comets.

When did the planets form? Using techniques introduced in Chapter 10, geologists have found that special types of meteorites thought to be leftover planetesimals formed at 4.57 Ga, and thus consider that date to be the birth date of the Solar System. If this date is correct, it means that the Solar System formed about 9 billion years after the Big Bang, and thus is only about a third as old as the Universe.

Differentiation of the Earth and Formation of the Moon

When planetesimals started to form, they had a fairly homogeneous distribution of material throughout, because the smaller pieces from which they formed all had much the same composition and collected together in no particular order. But large planetesimals did not stay homogeneous for long, because they began to heat up. The heat came primarily from three sources: the heat produced during collisions (similar to the phenomenon that happens when you bang on a nail with a hammer and they

both get warm), the heat produced when matter is squeezed into a smaller volume, and the heat produced from the decay of radioactive elements. In bodies whose temperature rose sufficiently to cause internal melting, denser iron alloy separated out and sank to the center of the body, whereas lighter rocky materials remained in a shell surrounding the center. By this process, called **differentiation**, protoplanets and large planetesimals developed internal layering early in their history. As we will see later, the central ball of iron alloy constitutes the body's core and the outer shell constitutes its mantle.

In the early days of the Solar System, planets continued to be bombarded by **meteorites** (solid objects, such as fragments of planetesimals, falling from space that land on a planet) even after the Sun had ignited and differentiation had occurred (**Box 1.2**). Heavy bombardment in the early days of the Solar System pulverized the surfaces of planets and eventually left huge numbers of craters (**See for Yourself A**). Bombardment also contributed to heating the planets.

Based on analysis and the dating of Moon rocks, most geologists have concluded that at about 4.53 Ga, a Mars-sized protoplanet slammed into the newborn Earth. In the process, the colliding body disintegrated and melted, along with a large part of the Earth's mantle. A ring of debris formed around the remaining, now-molten Earth, and quickly coalesced to form the Moon. Not all moons in the Solar System necessarily formed in this manner. Some may have been independent protoplanets or comets that were captured by a larger planet's gravity.

Why Are Planets Round?

Small planetesimals were jagged or irregular in shape, and asteroids today have irregular shapes. Planets, on the other hand, are more or less spherical. Why? Simply put, when a

> ### Did you ever wonder . . .
> is the Moon as old as the Earth?

FIGURE 1.7 The grainy interior of this meteorite may resemble the texture of a small planetesimal.

BOX 1.2   CONSIDER THIS ...

Meteors and Meteorites

During the early days of the Solar System, the Earth collided with and incorporated countless planetesimals and smaller fragments of solid material lying in its path. Intense bombardment ceased about 3.9 Ga, but even today collisions with space objects continue, and over 1,000 tons of material (rock, metal, dust, and ice) fall to Earth, on average, every year. The vast majority of this material consists of fragments derived from comets and asteroids sent careening into the path of the Earth after billiard-ball-like collisions with each other out in space, or because of the gravitational pull of a passing planet deflected their orbit. Some of the material, however, consists of chips of the Moon or Mars, ejected into space when large objects collided with those bodies.

Astronomers refer to any object from space that enters the Earth's atmosphere as a meteoroid. Meteoroids move at speeds of 20 to 75 km/s (over 45,000 mph), so fast that when they reach an altitude of about 150 km, friction with the atmosphere causes them to heat up and vaporize, leaving a streak of bright, glowing gas. The glowing streak, an atmospheric phenomenon, is a **meteor** (also known

colloquially, though incorrectly, as a "falling star") (**Fig. Bx1.2a**). Most visible meteors completely vaporize by an altitude of about 30 km. But dust-sized ones may slow down sufficiently to float to Earth, and larger ones (fist-sized or bigger) can survive the heat of entry to reach the surface of the planet. In some cases, meteoroids explode in brilliant fireballs.

Objects that strike the Earth are called **meteorites**. Although almost all meteorites are small and have not caused notable damage on Earth during human history, a very few have smashed through houses, dented cars, and bruised people. During the longer term of Earth history, however, there have been some catastrophic collisions that left huge craters (**Fig. Bx1.2b**).

Most meteorites are asteroidal or planetary fragments, for icy material is too fragile to survive the fall. Researchers recognize three basic classes of meteorites: iron (made of iron-nickel alloy), stony (made of rock), and stony iron (rock embedded in a matrix of metal). Of all known meteorites, about 93% are stony and 6% are iron (**Fig. Bx1.2c**). Researchers have concluded that some meteors (a special subcategory of stony meteorites

called carbonaceous chondrites, because they contain carbon and small spherical nodules called chondrules) are asteroids derived from planetesimals that never underwent differentiation into a core and mantle. Other stony meteorites and all iron meteorites are asteroids derived from planetesimals that had differentiated into a metallic core and a rocky mantle early in Solar System history but later shattered into fragments during collisions with other planetesimals. Most meteorites appear to be about 4.54 Ga, but carbonaceous chondrites are as old as 4.57 Ga and are the oldest solar system materials ever measured.

Since meteorites represent fragments of undifferentiated and differentiated planetesimals, geologists consider the *average* composition of meteorites to be representative of the *average* composition of the whole Earth. In other words, the estimates that geologists use for the proportions of different elements in the Earth are based largely on studying meteorites. Stony meteorites are probably similar in composition to the mantle, and iron meteorites are probably similar in composition to the core.

FIGURE Bx1.2 Meteors and meteorites.

(a) A shower of meteors over Hong Kong in 2001.

(b) The Barringer meteor crater in Arizona. It formed about 50,000 years ago and is 1.1 km in diameter.

The meteorite forming the crater was 50 m across.

(c) Examples of stony meteorites (left) and iron meteorites (right).

protoplanet gets big enough, gravity can change its shape. To picture how, imagine a block of cheese sitting outside on a hot summer day. As the cheese gets softer and softer, gravity causes it to spread out in a pancake-like blob. This model shows that gravitational force alone can cause material to change shape if the material is soft enough. Now let's apply this model to planetary growth.

The rock composing a small planetesimal is cool and strong enough so that the force of gravity is not sufficient to cause the rock to flow. But once a planetesimal grows beyond a certain critical size (about 1,000 km in diameter), its interior becomes warm and soft enough to flow in response to gravity. As a consequence, protrusions are pulled inward toward the center, and the planetesimal re-forms into a special shape that permits the force of gravity to be nearly the same at all points on its surface. This special shape is spherical because in a sphere mass is evenly distributed around the center (see Geology at a Glance, pp. 22–23).

> ## Take-Home Message
> Heavier elements formed in stars and supernovas added to gases in nebulae from which new generations of stars formed. Planets formed from rings of dust and ice orbiting the stars. As they formed, planets differentiated, with denser materials sinking to the center.

1.5 Welcome to the Neighborhood

Introducing the Earth System

So far in this chapter, we've described scientific ideas about how the Universe, and then the Solar System, formed. Now, let's focus on our home planet and develop an image of the Earth's overall architecture. We will see that our planet consists of several components—the atmosphere (Earth's gaseous envelope), the hydrosphere (Earth's surface and near-surface water), the biosphere (Earth's great variety of life forms), the lithosphere (the outer shell of the Earth), and the interior (material inside the Earth). These components, and the complex interactions among them, comprise the **Earth System**. Our planet remains a dynamic place. Heat retained inside from the planet's formation, as well as from continuing radioactive decay, provides the energy to move the lithosphere. And heat from the Sun keeps the atmosphere and hydrosphere in motion.

To get an initial sense of what the Earth System looks like, imagine that we are explorers from space rocketing toward the Earth for a first visit. After we pass the Moon, we begin to detect the Earth's magnetic field. Then we enter its atmosphere and orbit its surface.

A First Glance at the Magnetic Field

As our rocket approaches the Earth, and we see its beautiful bluish glow (**Fig. 1.8**), our instruments detect the planet's magnetic field, like a signpost shouting, "Approaching Earth!" A **magnetic field**, in a general sense, is the region affected by the force emanating from a magnet. In the context of physics, a force is a push or pull that can cause the velocity or shape of an object to change. Magnetic force, which grows progressively stronger as you approach the magnet, can attract or repel another magnet and can cause charged particles to move. Earth's magnetic field, like the familiar magnetic field around a bar magnet, is largely a **dipole**, meaning that it has two poles—a north pole and a south pole (**Fig. 1.9a, b**). When you bring two bar magnets close to one another, opposite poles attract and like poles repel. By convention, we represent the orientation of a magnetic dipole by an arrow that points from the south pole to the north pole, and we represent the magnetic field of a magnet by a set of invisible magnetic field lines that curve through the space around the magnet and enter the magnet at its poles. Arrowheads along these lines point in a direction to complete a loop. Magnetized needles, such as iron filings or compass needles, when placed in a field align with the magnetic field lines.

If we represent the Earth's magnetic field as emanating from an imaginary bar magnet in the planet's interior, the north pole of this bar lies near the south geographic pole of the Earth, whereas the south pole of the bar lies near the north geographic pole. (The *geographic poles* are the places where the spin axis of the Earth intersects the planet's surface.)

FIGURE 1.8 A view of the Earth as seen from space.

Forming the Planets and the Earth-Moon System

1. Forming the solar system, according to the nebular hypothesis: A nebula forms from hydrogen and helium left over from the Big Bang, as well as from heavier elements that were produced by fusion reactions in stars or during explosions of stars.

2. Gravity pulls gas and dust inward to form an accretion disk. Eventually a glowing ball—the proto-Sun—forms at the center of the disk.

6. Gravity reshapes the proto-Earth into a sphere. The interior of the Earth differentiates into a core and mantle.

5. Forming the planets from planetesimals: Planetesimals grow by continuous collisions. Gradually, an irregularly shaped proto-Earth develops. The interior heats up and becomes soft.

8. The Moon forms from the ring of debris.

7. Soon after Earth forms, a small planet collides with it, blasting debris that forms a ring around the Earth.

3. "Dust" (particles of refractory materials) concentrates in the inner rings, while "ice" (particles of volatile materials) concentrates in the outer rings. Eventually, the dense ball of gas at the center of the disk becomes hot enough for fusion reactions to begin. When it ignites, it becomes the Sun.

4. Dust and ice particles collide and stick together, forming planetesimals.

9. Eventually, the atmosphere develops from volcanic gases. When the Earth becomes cool enough, moisture condenses and rains to create the oceans. Some gases may be added by passing comets.

23

FIGURE 1.9   A magnetic field permeates the space around the Earth. It can be symbolized by a bar magnet.

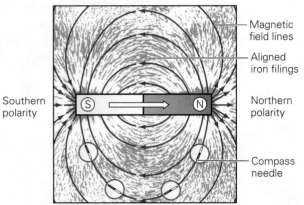

(a) A bar magnet produces a magnetic field. Magnetic field lines point into the "south pole" and out from the "north pole."

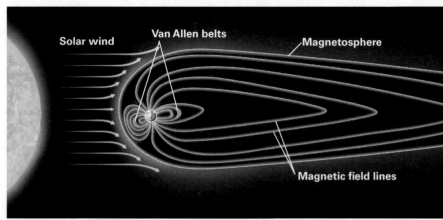

(c) Earth behaves like a magnetic dipole, but the field lines are distorted by the solar wind. The Van Allen radiation belts trap charged particles.

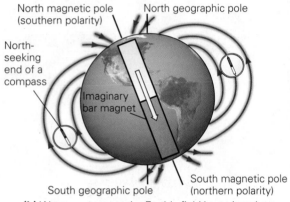

(b) We can represent the Earth's field by an imaginary bar magnet inside.

(d) Charged particles flow toward Earth's magnetic poles and cause gases in the atmosphere to glow, forming colorful aurorae in polar skies.

Nevertheless, geologists and geographers by convention refer to the magnetic pole closer to the north geographic pole as the north magnetic pole, and the magnetic pole closer to the south geographic pole as the south magnetic pole. This way, the north-seeking end of a compass points toward the north geographic pole, since opposite ends of a magnet attract.

Our Sun's stellar wind, known as the "solar wind," interacts with Earth's magnetic field, distorting it into a huge teardrop pointing away from the Sun. The solar wind consists of dangerous, high-velocity charged particles. Fortunately, the magnetic field deflects most (but not all) of the particles, so that they do not reach Earth's surface. In this way, the magnetic field acts like a shield against the solar wind; the region inside this magnetic shield is called the magnetosphere (**Fig. 1.9c**).

Though it protects the Earth from most of the solar wind, the magnetic field does not stop our spaceship, and we continue to speed toward the planet. At distances of about 3,000 km and 10,500 km out from the Earth, we encounter the Van Allen radiation belts, named for the physicist who first recognized them in 1959. These belts trap solar wind particles as well as cosmic rays (nuclei of atoms emitted from supernova explosions) that were moving so fast they were able to penetrate the weaker outer part of the magnetic field. Some charged particles make it past the Van Allen belts and are channeled along magnetic field lines to the polar regions of Earth. When these particles interact with gas atoms in the upper atmosphere, they cause the gases to glow, like the gases in neon signs, creating spectacular aurorae (**Fig. 1.9d**).

Introducing the Atmosphere

As our spaceship descends further, we enter Earth's **atmosphere**, an envelope of gas consisting of 78% nitrogen (N_2) and 21% oxygen (O_2), with minor amounts (1% total) of argon, carbon dioxide (CO_2), neon, methane, ozone, carbon monoxide, and sulfur dioxide (**Fig. 1.10a, b**). Other terrestrial planets have atmospheres, but they are not like Earth's.

The weight of overlying air squeezes down on the air below, pushing gas molecules closer together. Thus, both the density of air and the air pressure (the amount of push that the air exerts on material beneath it) increases closer to the surface (**Fig. 1.10c**). Technically, we specify pressure in units of force per unit area. Such units include atmospheres (abbreviated atm) and bars, where 1 atm = 1.04 kilograms per square centimeter, or 14.7 pounds per square inch. An atmosphere and a bar are almost the same: 1 atm = 1.01 bars. At sea level, average air pressure is 1 atm. Air pressure (and, therefore, density) decreases by half for every 5.6 km that you rise above sea level. Thus, at the peak of Mt. Everest, the highest point of the planet (8.85 km above sea level), air pressure is only 0.3 atm. People can't survive for long at elevations above about 5.5 km. Because of the decrease in density with elevation, 99% of atmospheric gas lies at elevations below 50 km, and the atmosphere is barely detectable at elevations above 120 km. The vacuum (lack of matter) characterizing interplanetary space lies above an elevation of about 600 km.

The nature of the atmosphere changes with increasing distance from the Earth's surface. Because of these changes, atmospheric scientists divide the atmosphere into layers. Most winds and clouds develop only in the lowest layer, the troposphere. The layers of the atmosphere that lie above the troposphere are named, in sequence from base to top: the stratosphere, the mesosphere, and the thermosphere (**Fig. 1.10d**). Boundaries between layers are defined as elevations at which temperature stops decreasing and starts increasing, or vice versa. Boundaries are named for the underlying layer. For example, the boundary between the troposphere and the overlying stratosphere is the tropopause.

> **Did you ever wonder . . .**
> how thick is our atmosphere?

FIGURE 1.10 Characteristics of the atmosphere that envelops the Earth.

(a) An orbiting astronaut's photograph shows the haze of the atmosphere fading up into the blackness of space.

(b) Composition of atmosphere. Nitrogen and oxygen dominate.

Nitrogen (N_2) 78.08%
Other gases (0.97%)
Oxygen (O_2) 20.95%

99.9997% of the atmosphere lies below an elevation of 100 km.

Record for balloon flight 34.7 km
Less dense (molecules far apart)
F-22 Raptor 19 km
Commercial jet 12–15 km
Gravity
More dense (molecules close together)
Mt. Everest 8,848 m
Denali 6,189 m
Mauna Kea 4,205 m
Cirrus clouds
Altitude (km)
Pressure (bars)

Meteor
Thermosphere
Mesopause
Mesosphere
Temperature gradient
Stratopause
Ozone interval
Stratosphere
Tropopause
Troposphere
Altitude (km)
Temperature
−100° −80° −60° −40° −20° 0° 20° 40°C
−160° −120° −80° −40° 0° 32° 60° 100°F

(c) Molecules pack together more tightly at the base of the atmosphere, so atmospheric pressure changes with elevation.

(d) The atmosphere can be divided into several distinct layers. We live in the troposphere.

Land and Oceans

Imagine that we now guide our spaceship into orbit around the Earth. Our first task is to make a map of the planet. What features should go on this map? Starting with the most obvious, we note that land (continents and islands) covers about 30% of the surface (**Fig. 1.11**). Some of the land surface consists of solid rock, whereas some has a covering of **sediment** (materials such as sand and gravel, in which the grains are not stuck together). The amount of cover by vegetation varies widely (See For Yourself A). Surface water covers the remaining 70% of the Earth. Most surface water is salty and in oceans, but some is fresh and fills lakes and rivers. Our instruments also detect groundwater, the water that fills cracks and holes (pores) within rock and sediment under the land surface. Finally, we find that ice covers significant areas of land and sea in polar regions and at high elevations, and that living organisms populate the land, sea, air, and even the upper few kilometers of the subsurface.

To finish off our map of the Earth's surface, we note that the surface is not flat. Topography, the variation in elevation of the land surface, defines plains, mountains, and valleys (See for Yourself A). Similarly, **bathymetry**, the variation in elevation of the ocean floor, defines mid-ocean ridges, abyssal plains, and deep-ocean trenches (see Fig. 1.11). The deepest point on the ocean floor is 10.9 km below sea level, and the highest point on land is almost 8.9 km above—the total difference in elevation (19.8 km) is only 0.3% of Earth's radius (6,371 km).

1.6 Looking Inward—Introducing the Earth's Interior

What Is the Earth Made Of?

At this point, we leave our fantasy space voyage and turn our attention inward to the materials that make up the solid Earth, because we need to be aware of these before we can discuss the architecture of the Earth's interior. Let's begin by reiterating that the Earth consists mostly of elements produced by fusion reactions in stars and by supernova explosions. Only four elements (iron, oxygen, silicon, and magnesium) make up 91.2% of the Earth's mass; the remaining 8.8% consists of the other 88 elements (**Fig. 1.12**). The elements of the Earth comprise a great variety of materials. For reference in this chapter and the next, we introduce the basic categories of materials. All of these will be discussed in more detail later in the book.

> *Organic chemicals*. Carbon-containing compounds that either

FIGURE 1.11 This map of the Earth shows variations in elevation on both the land surface and the sea floor. Darker blues are deeper water in the ocean. Greens are lower elevation on land.

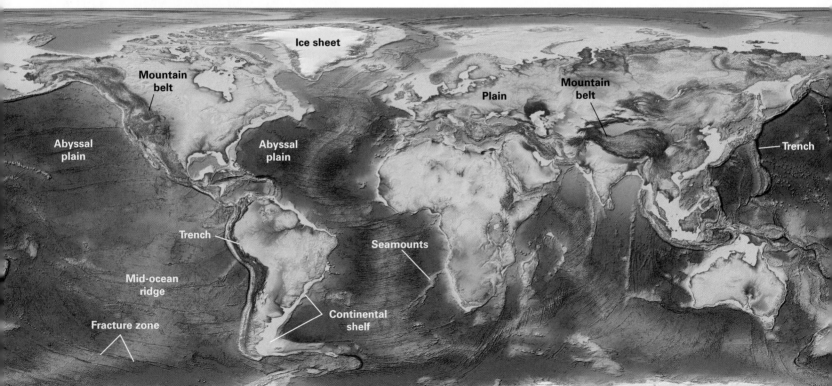

FIGURE 1.12 The proportions of major elements making up the mass of the whole Earth.

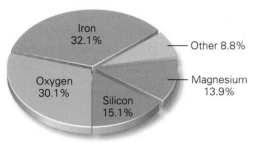

felsic (or *silicic*), *intermediate*, *mafic*, and *ultramafic*. As the proportion of silica in a rock increases, the density (mass per unit volume) decreases. Thus, felsic rocks are less dense than mafic rocks. Many different rock types occur in each class, as will be discussed in detail in Chapters 4 through 7. For now, we introduce the four rock types whose names we need to know for our discussion of the Earth's layers that follows. These are (1) *granite*, a felsic rock with large grains; (2) *gabbro*, a mafic rock with large grains; (3) *basalt*, a mafic rock with small grains; and (4) *peridotite*, an ultramafic rock with large grains.

occur in living organisms or have characteristics that resemble compounds in living organisms are called *organic chemicals*.

> *Minerals.* A solid, natural substance in which atoms are arranged in an orderly pattern is a **mineral**. A single coherent sample of a mineral that grew to its present shape is a crystal, whereas an irregularly shaped sample, or a fragment derived from a once-larger crystal or cluster of crystals, is a grain.

> *Glasses.* A solid in which atoms are not arranged in an orderly pattern is called *glass*.

> *Rocks.* Aggregates of mineral crystals or grains, or masses of natural glass, are called *rocks*. Geologists recognize three main groups of rocks. (1) *Igneous* rocks develop when hot molten (liquid) rock cools and freezes solid. (2) *Sedimentary* rocks form from grains that break off preexisting rock and become cemented together, or from minerals that precipitate out of a water solution. (3) *Metamorphic* rocks form when preexisting rocks change in response to heat and pressure.

> *Sediment.* An accumulation of loose mineral grains (grains that have not stuck together) is called *sediment*.

> *Metals.* A solid composed of metal atoms (such as iron, aluminum, copper, and tin) is called a **metal**. An **alloy** is a mixture containing more than one type of metal atom.

> *Melts.* A **melt** forms when solid materials become hot and transform into liquid. Molten rock is a type of melt—geologists distinguish between *magma*, which is molten rock beneath the Earth's surface, and *lava*, molten rock that has flowed out onto the Earth's surface.

> *Volatiles.* Materials that easily transform into gas at the relatively low temperatures found at the Earth's surface are called **volatiles**.

The most common minerals in the Earth contain **silica** (a compound of silicon and oxygen) mixed in varying proportions with other elements. These minerals are called silicate minerals. Not surprisingly, rocks composed of silicate minerals are *silicate rocks*. Geologists distinguish four classes of igneous silicate rocks based, in essence, on the proportion of silica to iron and magnesium. In order, from greatest to least proportion of silica to iron and magnesium, these classes are

Discovering the Earth's Internal Layers

People have speculated about what's inside our planet since ancient times. What is the source of incandescent lavas that spew from volcanoes, of precious gems and metals found in mines, of sparkling mineral waters that bubble from springs, and of the mysterious forces that shake the ground and topple buildings? In ancient Greece and Rome, the subsurface was the underworld, Hades, home of the dead, a region of fire and sulfurous fumes. Perhaps this image was inspired by the molten rock and smoke emitted by the volcanoes of the Mediterranean region. In the 18th and 19th centuries, European writers thought the Earth's interior resembled a sponge, containing open caverns variously filled with molten rock, water, or air. In fact, in the popular 1864 novel *Journey to the Center of the Earth*, by the French author Jules Verne, three explorers hike through interconnected caverns down to the Earth's center.

How can we explore the interior for real? We can't dig or drill down very far. Indeed, the deepest mine penetrates only about 3.5 km beneath the surface of South Africa. And the deepest drill hole probes only 12 km below the surface of northern Russia—compared with the 6,371 km radius of the Earth, this hole makes it less than 0.2% of the way to the center and is nothing more than a pinprick. Our modern image of the Earth's interior, one made up of distinct layers, is the end product of many discoveries made during the past 200 years.

The first clue that led away from Jules Verne's sponge image came when researchers successfully measured the mass of the whole Earth, and from this information derived its *average* density. They found that the average density of our planet far exceeds the density of common rocks found on the surface. Thus, the interior of the Earth must contain denser material than its outermost layer and can't possibly be full of holes. In fact, the mass of the Earth overall is so great that the planet must contain a large amount of metal. Since the Earth is close to being a sphere, the metal must be concentrated near the center. Otherwise, centrifugal force due to the spin of the Earth on its axis would pull the equator out, and the

FIGURE 1.13 An early image of Earth's internal layers.

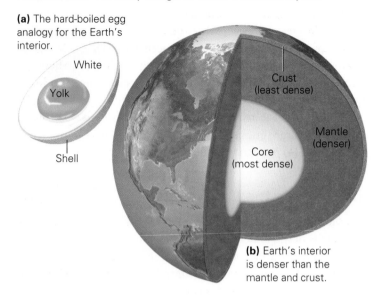

(a) The hard-boiled egg analogy for the Earth's interior.

White

Yolk

Shell

Crust (least dense)

Mantle (denser)

Core (most dense)

(b) Earth's interior is denser than the mantle and crust.

planet would become a disk. (To picture why, consider that when you swing a hammer, your hand feels more force if you hold the end of the light wooden shaft, rather than the heavy metal head.) Finally, researchers realized that, though molten rock occasionally oozes out of the interior at volcanoes, the interior must be mostly solid, because if it weren't, the land surface would rise and fall due to tidal forces much more than it does.

Eventually, researchers concluded that the Earth resembled a hard-boiled egg, in that it had three principal layers: a not-so-dense **crust** (like an eggshell, composed of rocks such as granite, basalt, and gabbro), a denser solid **mantle** in the middle (the "white," composed of a then-unknown material), and a very dense **core** (the "yolk," composed of an unknown metal) (**Fig. 1.13a, b**). Clearly, many questions remained. How thick are the layers? Are the boundaries between layers sharp or gradational? And what exactly are the layers composed of?

Clues from the Study of Earthquakes: Refining the Image

When rock within the outer portion of the Earth suddenly breaks and slips along a fracture called a *fault*, it generates shock waves (abrupt vibrations), called seismic waves, that travel through the surrounding rock outward from the break. Where these waves cause the surface of the Earth to vibrate, people feel an **earthquake**, an episode of ground shaking. You can simulate this process, at a small scale, when you break a stick between your hands and feel the snap with your hands (**Fig. 1.14a, b**).

In the late 19th century, geologists learned that earthquake energy could travel, in the form of waves, all the way through the Earth's interior from one side to the other. Geologists immediately realized that the study of earthquake waves traveling through the Earth might provide a tool for exploring the Earth's insides, much as ultrasound today helps doctors study a patient's insides. Specifically, laboratory measurements demonstrated that earthquake waves travel at different velocities (speeds) through different materials. Thus, by detecting depths at which velocities suddenly change, geoscientists pinpointed the boundaries between layers and even recognized subtler boundaries within layers. For example, such studies led geoscientists to subdivide the mantle into the upper mantle and lower mantle, and subdivide the core into the inner core and outer core. (Chapter 8 provides further details about earthquakes, and Interlude D shows how the study of earthquake waves defines the Earth's layers.)

Pressure and Temperature Inside the Earth

In order to keep underground tunnels from collapsing under the pressure created by the weight of overlying rock, mining engineers must design sturdy support structures. It is no surprise

FIGURE 1.14 Faulting and earthquakes.

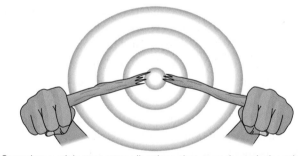

(a) Snapping a stick generates vibrations that pass through the stick to your hands.

Earthquake wave

Fault plane

(not to scale)

(b) Similarly, when the rock inside the Earth suddenly breaks and slips, forming a fracture called a fault, it generates shock waves that pass through the Earth and shake the surface.

that deeper tunnels require stronger supports: the downward push from the weight of overlying rock increases with depth, simply because the mass of the overlying rock layer increases with depth. In solid rock, the pressure at a depth of 1 km is about 300 atm. At the Earth's center, pressure probably reaches about 3,600,000 atm.

Temperature also increases with depth in the Earth. Even on a cool winter's day, miners who chisel away at gold veins exposed in tunnels 3.5 km below the surface swelter in temperatures of about 53°C (127°F). We refer to the rate of change in temperature with depth as the **geothermal gradient**. In the upper part of the crust, the geothermal gradient averages between 20°C and 30°C per km. At greater depths, the rate decreases to 10°C per km or less. Thus, 35 km below the surface of a continent, the temperature reaches 400°C to 700°C, and the mantle-core boundary is about 3,500°C. No one has ever directly measured the temperature at the Earth's center, but calculations suggest it may exceed 4,700°C, close to the Sun's surface temperature of 5,500°C.

> **Did you ever wonder...**
> how hot it gets at the center of this planet ?

> **Take-Home Message**
> The Earth consists of many materials, the most common of which is silicate rock. Studies, including analysis of earthquake waves, show that the Earth can be divided into three layers—the crust, the mantle, and the core. Temperature and pressure increase with depth.

1.7 What Are the Layers Made Of?

As a result of studies during the past century, geologists have a pretty clear sense of what the layers inside the Earth are made of. Let's now look at the properties of individual layers in more detail (**Fig. 1.15a, b**).

The Crust

When you stand on the surface of the Earth, you are standing on top of its outermost layer, the crust. The crust is our home and the source of all our resources. How thick is this all-important layer? Or, in other words, what is the depth to the crust-mantle boundary? An answer came from the studies of Andrija Mohorovičić, a researcher working in Zagreb, Croatia. In 1909, he discovered that the velocity of earthquake waves

suddenly increased at a depth of tens of kilometers beneath the Earth's surface, and he suggested that this increase was caused by an abrupt change in the properties of rock (see Interlude D for further details). Later studies showed that this change can be found most everywhere around our planet, though it occurs at different depths in different locations. Specifically, it's deeper beneath continents than beneath oceans. Geologists now consider the change to define the base of the crust, and they refer to it as the **Moho** in Mohorovičić's honor. The relatively shallow depth of the Moho (7 to 70 km, depending on location) as compared to the radius of the Earth (6,371 km) emphasizes that the crust is very thin indeed. In fact, the crust is only about 0.1% to 1.0% of the Earth's radius, so if the Earth were the size of a balloon, the crust would be about the thickness of the balloon's skin.

The crust is not simply cooled mantle, like the skin on chocolate pudding, but rather consists of a variety of rocks that differ in composition (chemical makeup) from mantle rock. Geologists distinguish between two fundamentally different types of crust—oceanic crust, which underlies the sea floor, and continental crust, which underlies continents.

Oceanic crust is only 7 to 10 km thick. At highway speeds (100 km per hour), you could drive a distance equal to the thickness of the oceanic crust in about five minutes. At the top, we find a blanket of sediment, generally less than 1 km thick, composed of clay and tiny shells that settled like snow out of the sea. Beneath this blanket, the oceanic crust consists of a layer of basalt and, below that, a layer of gabbro.

Most continental crust is about 35 to 40 km thick—about four to five times the thickness of oceanic crust—but its thickness varies significantly. In some places, continental crust has been stretched and thinned so it's only 25 km from the surface to the Moho, and in some places, the crust has been crumpled and thickened to become up to 70 km thick. In contrast to oceanic crust, continental crust contains a great variety of rock types, ranging from mafic to felsic in composition. On average, upper continental crust is less mafic than oceanic crust—it has a felsic (granite-like) to intermediate composition—so continental crust overall is less dense than oceanic crust. Notably, oxygen is the most abundant element in the crust (**Fig. 1.16**).

The Mantle

The mantle of the Earth forms a 2,885-km-thick layer surrounding the core. In terms of volume, it is the largest part of the Earth. In contrast to the crust, the mantle consists entirely of an ultramafic (dark and dense) rock called peridotite. This means that peridotite, though rare at the Earth's surface, is actually the most abundant rock in our planet! Researchers have found that earthquake-wave velocity changes at a depth

FIGURE 1.15 A modern view of Earth's interior layers.

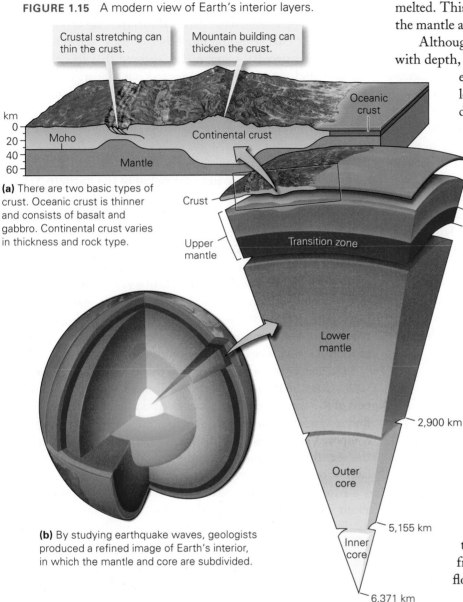

Crustal stretching can thin the crust.

Mountain building can thicken the crust.

(a) There are two basic types of crust. Oceanic crust is thinner and consists of basalt and gabbro. Continental crust varies in thickness and rock type.

(b) By studying earthquake waves, geologists produced a refined image of Earth's interior, in which the mantle and core are subdivided.

melted. This melt occurs in films or bubbles between grains in the mantle at a depth of 100 to 200 km beneath the ocean floor.

Although overall, the temperature of the mantle increases with depth, temperature also varies significantly with location even at the same depth. The warmer regions are less dense, while the cooler regions are denser. The distribution of warmer and cooler mantle indicates that the mantle convects like water in a simmering pot; warmer mantle is relatively buoyant and gradually flows upward, while cooler, denser mantle sinks.

The Core

Early calculations suggested that the core had the same density as gold, so for many years people held the fanciful hope that vast riches lay at the heart of our planet. Alas, geologists eventually concluded that the core consists of a far less glamorous material, iron alloy (iron mixed with tiny amounts of other elements). Studies of seismic waves led geoscientists to divide the core into two parts, the outer core (between 2,900 and 5,155 km deep) and the inner core (from a depth of 5,155 km down to the Earth's center at 6,371 km). The outer core consists of *liquid* iron alloy. It can exist as a liquid because the temperature in the outer core is so high that even the great pressures squeezing the region cannot keep atoms locked into a solid framework. The iron alloy of the outer core can flow, and this flow generates Earth's magnetic field.

The inner core, with a radius of about 1,220 km, is a *solid* iron alloy that may reach a temperature of over 4,700°C. Even though it is hotter than the outer core, the inner core is a solid because it is deeper and is subjected to even greater pressure. The pressure keeps atoms locked together tightly in very dense materials.

of 400 km and again at a depth of 660 km in the mantle. Based on this observation, they divide the mantle into two sublayers: the **upper mantle**, down to a depth of 660 km, and the **lower mantle**, from 660 km down to 2,900 km. The **transition zone** is the interval between 400 km and 660 km deep.

Almost all of the mantle is solid rock. But even though it's solid, mantle rock below a depth of 100 to 150 km is so hot that it's soft enough to flow. This flow, however, takes place extremely slowly—at a rate of less than 15 cm a year. *Soft* here does not mean liquid; it simply means that over long periods of time mantle rock can change shape without breaking. We stated earlier that *almost* all of the mantle is solid. We used the word "almost" because up to a few percent of the mantle has

The Lithosphere and the Asthenosphere

So far, we have identified three major layers (crust, mantle, and core) inside the Earth that differ compositionally from each other. Earthquake waves travel at different velocities through these layers. An alternative way of thinking about Earth layers comes from studying the degree to which the material making up a layer can flow. In this context, we distinguish between *rigid* materials, which can bend or break but cannot flow, and *plastic* materials, which are relatively soft and can flow without breaking.

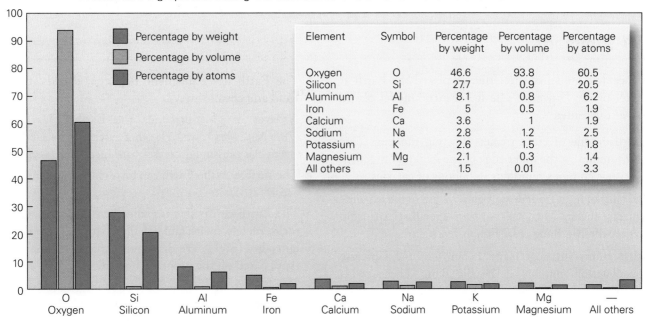

FIGURE 1.16 A table and a graph illustrating the abundance of elements in the Earth's crust.

Element	Symbol	Percentage by weight	Percentage by volume	Percentage by atoms
Oxygen	O	46.6	93.8	60.5
Silicon	Si	27.7	0.9	20.5
Aluminum	Al	8.1	0.8	6.2
Iron	Fe	5	0.5	1.9
Calcium	Ca	3.6	1	1.9
Sodium	Na	2.8	1.2	2.5
Potassium	K	2.6	1.5	1.8
Magnesium	Mg	2.1	0.3	1.4
All others	—	1.5	0.01	3.3

Geologists have determined that the outer 100 to 150 km of the Earth is relatively rigid. In other words, the Earth has an outer shell composed of rock that cannot flow easily. This outer layer is called the **lithosphere**, and it consists of the crust plus the uppermost, cooler part of the mantle. We refer to the portion of the mantle within the lithosphere as the lithospheric mantle. Note that the terms lithosphere and crust are not synonymous—the crust is just the upper part of the lithosphere. The lithosphere lies on top of the **asthenosphere**, which is the portion of the mantle in which rock can flow. The boundary between the lithosphere and asthenosphere occurs where the temperature reaches about 1280°C, for at temperatures higher than this value mantle rock becomes soft enough to flow.

Geologists distinguish between two types of lithosphere (**Fig. 1.17**). Oceanic lithosphere, topped by oceanic crust, generally has a thickness of about 100 km. In contrast, continental lithosphere, topped by continental crust, generally has a thickness of about 150 km. Notice that the

asthenosphere is entirely in the mantle and generally lies below a depth of 100 to 150 km. We can't assign a specific depth to the base of the asthenosphere because all of the mantle below 150 km can flow, but for convenience, some geologists consider the base of the asthenosphere to be the top of the transition zone.

Now, with an understanding of Earth's overall architecture at hand, we can discuss geology's grand unifying theory—plate tectonics. The next chapter introduces this key topic.

Take-Home Message

Earth's outermost layer, the crust, is very thin; oceanic and continental crust differ in composition. Most of Earth's mass lies in the mantle. A metallic core lies at this planet's center. The crust and the outermost mantle together comprise the rigid lithosphere.

FIGURE 1.17 A block diagram of the lithosphere, emphasizing the difference between continental and oceanic lithosphere.

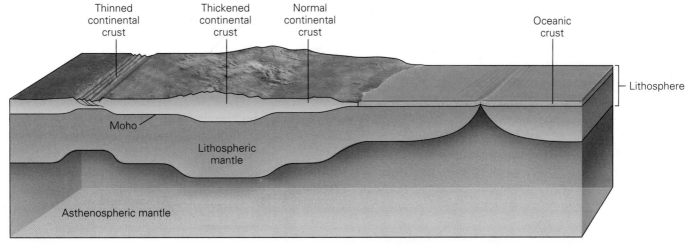

Chapter Summary

> The geocentric model of the Universe placed the Earth at the center of the Universe. The heliocentric model placed the Sun at the center.

> The Earth is one of eight planets orbiting the Sun. This Solar System lies on the outer edge of the Milky Way galaxy. The Universe contains hundreds of billions of galaxies.

> The red shift of light from distant galaxies led to the expanding Universe theory. Most astronomers agree that expansion began after the Big Bang, 13.7 billion years ago.

> The first atoms (hydrogen and helium) of the Universe developed within minutes of the Big Bang. These atoms formed vast gas clouds, called nebulae.

> Gravity caused clumps of gas in the nebulae to coalesce into flattened disks with bulbous centers. The center of each disk became so dense and hot that fusion reactions began and they became true stars.

> The Earth, and the life forms on it, contains elements that could have been produced only during the life cycle of stars. Thus, we are all made of stardust.

> Planets developed from the rings of gas and dust surrounding protostars. These condensed into planetesimals that then clumped together to form protoplanets, and finally true planets. Inner rings became the terrestrial planets. Outer rings grew into gas-giant planets.

> The Moon formed from debris ejected when a protoplanet collided with the Earth in the young Solar System.

> A planet assumes a near-spherical shape when it becomes so soft that gravity can smooth out irregularities.

> The Earth has a magnetic field that shields it from solar wind and cosmic rays.

> A layer of gas surrounds the Earth. This atmosphere (78% N_2, 21% O_2, and 1% other gases) can be subdivided into layers. Air pressure decreases with increasing elevation.

> The surface of the Earth can be divided into land (30%) and ocean (70%).

> The Earth consists of organic chemicals, minerals, glasses, rocks, metals, melts, and volatiles. Most rocks on Earth contain silica (SiO_2). We distinguish among various major rock types based on the proportion of silica.

> The Earth's interior can be divided into three distinct layers: the very thin crust, the rocky mantle, and the metallic core.

> Pressure and temperature both increase with depth in the Earth. The rate at which temperature increases as depth increases is the geothermal gradient.

> The crust is a thin skin that varies in thickness from 7–10 km (beneath the oceans) to 25–70 km (beneath the continents). Oceanic crust is mafic in composition, whereas average upper continental crust is felsic to intermediate. The mantle is composed of ultramafic rock. The core is made of iron alloy.

> Studies of earthquake waves reveal that the mantle can be subdivided into an upper mantle (including the transition zone) and a lower mantle. The core can be subdivided into the liquid outer core and a solid inner core. Circulation of the outer core produces the Earth's magnetic field.

> The crust plus the upper part of the mantle constitute the lithosphere, a rigid shell. The lithosphere lies over the asthenosphere, mantle that can flow.

Key Terms

alloy (p. 27)
asthenosphere (p. 31)
atmosphere (p. 24)
bathymetry (p. 26)
Big Bang theory (p. 15)
core (p. 28)
cosmology (p. 9)
crust (p. 28)
differentiation (p. 19)
dipole (p. 21)
Doppler effect (p. 13)
earthquake (p. 28)
Earth System (p. 21)
energy (p. 11)
expanding Universe theory (p. 15)

fission (p. 16)
frequency (p. 13)
fusion (p. 16)
galaxy (p. 11)
geocentric model (p. 10)
geothermal gradient (p. 29)
giant planet (p. 12)
gravity (p. 10)
heliocentric model (p. 10)
lithosphere (p. 31)
lower mantle (p. 30)
magnetic field (p. 21)
mantle (p. 28)
melt (p. 27)
metal (p. 27)

meteor (p. 20)
meteorite (pp. 19, 20)
mineral (p. 27)
Moho (p. 29)
moon (p. 13)
nebula (p. 17)
nebular theory (p. 18)
planet (p. 11)
planetesimal (p. 19)
protoplanetary disk (p. 18)
protoplanet (p. 19)
protostar (p. 17)
radioactive element (p. 16)
red shift (p. 13)
refractory (p. 18)

sediment (p. 26)
silica (p. 27)
Solar System (p. 11)
star (p. 11)
stellar nucleosynthesis (p. 17)
stellar wind (p. 17)
supernova (p. 17)
terrestrial planet (p. 12)
transition zone (p. 30)
Universe (p. 9)
upper mantle (p. 30)
volatile (pp. 18, 27)
wave (p. 13)
wavelength (p. 13)

Review Questions

1. Contrast the geocentric and heliocentric Universe concepts.

2. Describe how the Doppler effect works.

3. What does the red shift of the galaxies tell us about their motion with respect to the Earth?

4. What is the Big Bang, and when did it occur?

5. Describe the steps in the formation of the Solar System according to the nebular theory.

6. Why isn't the Earth homogeneous?

7. Describe how the Moon was formed.

8. Why is the Earth round?

9. What is the Earth's magnetic field? Draw a representation of the field on a piece of paper. What causes aurorae?

10. What is the Earth's atmosphere composed of? Why would you die of suffocation if you were to eject from a fighter plane at an elevation of 12 km without an oxygen tank?

11. What is the proportion of land area to sea area on Earth?

12. Describe the major categories of materials constituting the Earth. On what basis do geologists distinguish among different kinds of silicate rock?

13. What are the principal layers of the Earth?

14. How do temperature and pressure change with increasing depth in the Earth?

15. What is the Moho? Describe the differences between continental crust and oceanic crust.

16. What is the mantle composed of? Is there any melt in it?

17. What is the core composed of? How do the inner and outer cores differ? Which produces the magnetic field?

18. What is the difference between the lithosphere and asthenosphere? At what depth does the lithosphere-asthenosphere boundary occur? Is this above or below the Moho? In the asthenosphere entirely liquid?

On Further Thought

19. The further a galaxy lies from Earth, the greater its red shift. Why? (*Hint:* Draw two points that are initially 1 cm apart, and two points that are initially 2 cm apart. Imagine doubling the distance between the points in each pair in a given time.)

20. Did all first-generation stars form at the same time?

21. Why are the giant planets, which contain abundant gas and ice, further from the Sun?

22. Recent observations suggest that the Moon has a very small, solid core that is less than 3% of its mass. In comparison, Earth's core is about 33% of its mass. Explain why this difference might exist.

23. The Moon has virtually no magnetosphere. Why?

24. Popular media sometimes imply that the crust floats on a "sea of magma." Is this a correct image of the mantle just below the Moho? Explain your answer.

SEE FOR YOURSELF A...   Earth and Sky

Download *Google Earth*™ from the Web in order to visit the locations described below (instructions appear in the Preface of this book). You'll find further locations and associated active-learning exercises on Worksheet A of our **Geotours Workbook**.

Barringer Crater, Arizona
Latitude   35°1′35.23″N,
Longitude   111°1′22.23″W

This structure, also known as Meteor Crater, is 1.2 km in diameter and formed from the impact of an iron meteorite about 50,000 years ago. Viewed from an elevation of 5 km, you can see the raised rim of the crater.

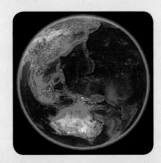

View of the Whole Earth
Latitude   05°56′59.21″N,
Longitude   143°6′23.83″E

An astronaut looking down on the western Pacific region from 14,000 km out in space can distinguish land from sea and can see the variation in amount of vegetation. Instruments reveal the shape of the sea-floor.

CHAPTER 2

The Way the Earth Works: Plate Tectonics

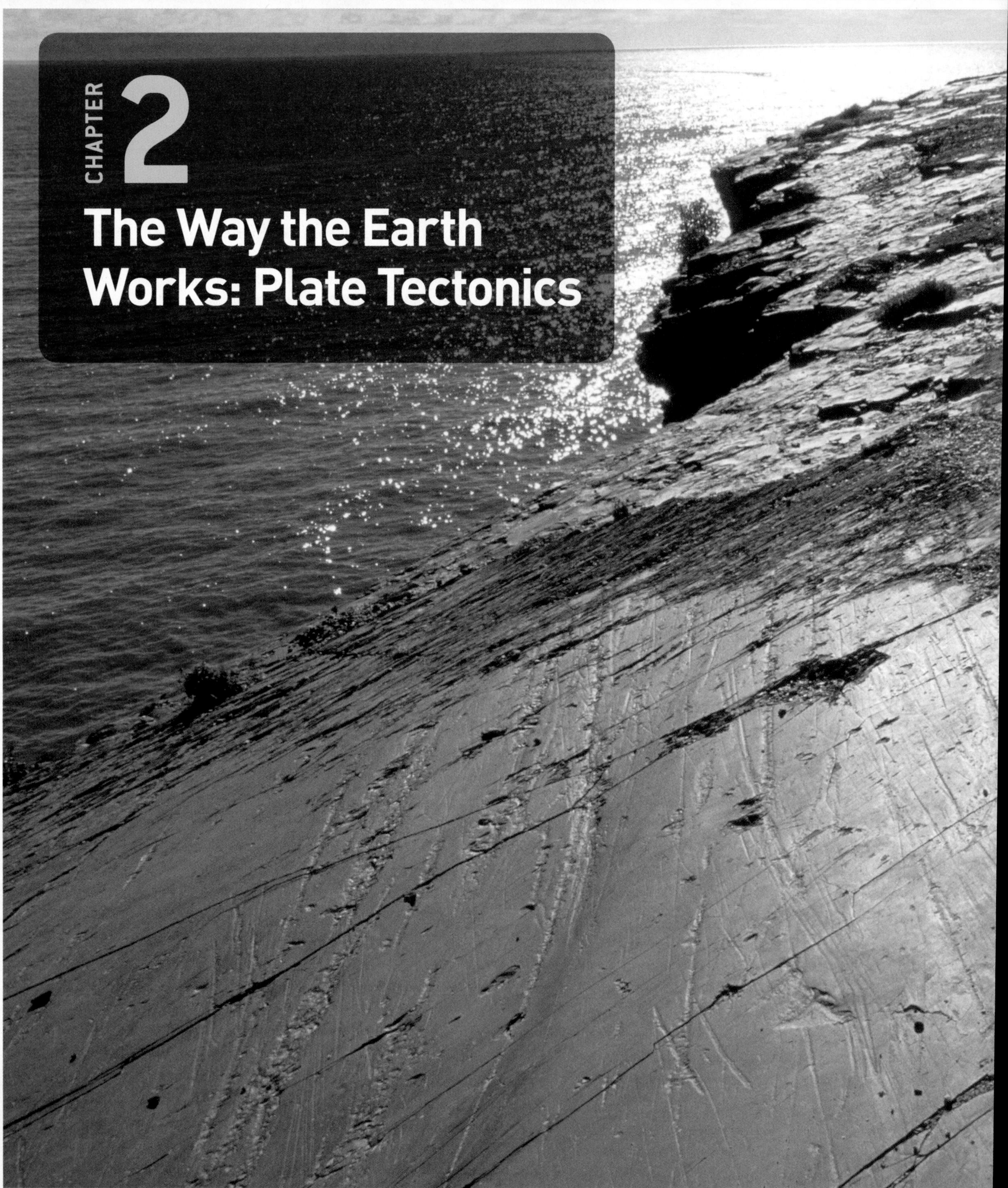

Chapter Objectives

By the end of this chapter you should know . . .

> Wegener's evidence for continental drift.

> how study of paleomagnetism proves that continents move.

> how sea-floor spreading works, and how geologists can prove that it takes place.

> that the Earth's lithosphere is divided into about 20 plates that move relative to one another.

> the three kinds of plate boundaries and the basis for recognizing them.

> how fast plates move, and how we can measure the rate of movement.

We are like a judge confronted by a defendant who declines to answer, and we must determine the truth from the circumstantial evidence.

—Alfred Wegener (German scientist, 1880–1930; on the challenge of studying the Earth)

2.1 Introduction

In September 1930, fifteen explorers led by a German meteorologist, Alfred Wegener, set out across the endless snowfields of Greenland to resupply two weather observers stranded at a remote camp. The observers had been planning to spend the long polar night recording wind speeds and temperatures on Greenland's polar plateau. At the time, Wegener was well known, not only to researchers studying climate but also to geologists. Some fifteen years earlier, he had published a small book, *The Origin of the Continents and Oceans*, in which he had dared to challenge geologists' long-held assumption that the continents had remained fixed in position through all of Earth history. Wegener thought, instead, that the continents once fit together like pieces of a giant jigsaw puzzle, to make one vast supercontinent. He suggested that this supercontinent, which he named **Pangaea** (pronounced pan-Jee-ah; Greek for all land), later fragmented into separate continents that drifted apart, moving slowly to their present positions (**Fig. 2.1a, b**). This model came to be known as **continental drift**.

Wegener presented many observations in favor of continental drift, but he met with strong resistance. At a widely publicized 1926 geology conference in New York City, a crowd of celebrated American professors scoffed, "What force could possibly be great enough to move the immense mass of a continent?" Wegener's writings didn't provide a good answer, so despite all the supporting observations he had provided, most of the meeting's participants rejected continental drift.

Now, four years later, Wegener faced his greatest challenge. On October 30, 1930, Wegener reached the observers and dropped off enough supplies to last the winter. Wegener and one companion set out on the return trip the next day, but they never made it home.

Had Wegener survived to old age, he would have seen his hypothesis become the foundation of a scientific revolution. Today, geologists accept many aspects of Wegener's ideas and take for granted that the map of the Earth constantly changes; continents waltz around this planet's surface, variously combining and breaking apart through geologic time. The revolution began in 1960, when an American geologist, Harry Hess, proposed that as continents drift apart, new ocean floor forms between them by a process that his contemporary, Robert Dietz, also had described and named **sea-floor spreading**. Hess and others suggested that continents move toward each other when the old ocean floor between them sinks back down into the Earth's interior, a process now called **subduction**. By 1968, geologists had developed a fairly complete model encompassing continental drift, sea-floor spreading, and subduction. In this model, Earth's lithosphere, its outer, relatively rigid shell, consists of about twenty distinct pieces, or plates, that slowly move relative to each other. Because we can confirm this model using many observations, it has gained the status of a theory, which we now call the theory of plate tectonics, or simply **plate tectonics**, from the Greek word *tekton*, which means builder; plate movements "build" regional geologic features. Geologists view plate tectonics as the grand unifying theory of geology, because it can successfully explain a great many geologic phenomena.

In this chapter, we introduce the observations that led Wegener to propose continental drift. Then we look at paleomagnetism, the record of Earth's magnetic field in the past, because it provides a key proof of continental drift. Next, we learn how observations about the sea floor, made by geologists during the mid-twentieth century, led Harry Hess to propose the concept of sea-floor spreading. We conclude by describing the many facets of modern plate tectonics theory.

2.2 Wegener's Evidence for Continental Drift

Wegener suggested that a vast supercontinent, Pangaea, existed until near the end of the Mesozoic Era (the interval of geologic time that lasted from 251 to 65 million years ago). He

FIGURE 2.1 Alfred Wegener and his model of continental drift.

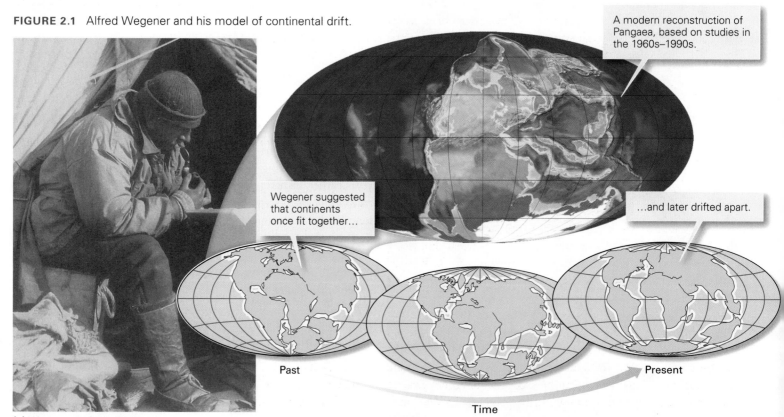

A modern reconstruction of Pangaea, based on studies in the 1960s–1990s.

Wegener suggested that continents once fit together...

...and later drifted apart.

Past   Present

Time

(a) Wegener in Greenland

(b) Wegener's maps illustrating continental drift as he drew them about 1915. Many details of the reconstruction have changed since, as illustrated by the inset.

suggested that Pangaea then broke apart, and the landmasses moved away from each other to form the continents we see today. Let's look at some of Wegener's arguments and see what led him to formulate this hypothesis of continental drift.

The Fit of the Continents

Almost as soon as maps of the Atlantic coastlines became available in the 1500s, scholars noticed the fit of the continents. The northwestern coast of Africa could tuck in against the eastern coast of North America, and the bulge of eastern South America could nestle cozily into the indentation of southwestern Africa. Australia, Antarctica, and India could all connect to the southeast of Africa, while Greenland, Europe, and Asia could pack against the northeastern margin of North America. In fact, all the continents could be joined, with remarkably few overlaps or gaps, to create Pangaea. Wegener concluded that the fit was too good to be coincidence and thus that the continents once *did* fit together.

> **Did you ever wonder...**
> why opposite coasts of the Atlantic look like they fit together?

Locations of Past Glaciations

Glaciers are rivers or sheets of ice that flow across the land surface. As a glacier flows, it carries sediment grains of all sizes (clay, silt, sand, pebbles, and boulders). Grains protruding from the base of the moving ice carve scratches, called striations, into the substrate. When the ice melts, it leaves the sediment in a deposit called till, that buries striations. Thus, the occurrence of till and striations at a location serve as evidence that the region was covered by a glacier in the past (see chapter opening photo). By studying the age of glacial till deposits, geologists have determined that large areas of the land were covered by glaciers during time intervals of Earth history called ice ages. One of these ice ages occurred from about 326 to 267 Ma, near the end of the Paleozoic Era.

Wegener was an Arctic climate scientist by training, so it's no surprise that he had a strong interest in glaciers. He knew that glaciers form mostly at high latitudes today. So he suspected that if he plotted a map of the locations of late Paleozoic glacial till and striations, he might gain insight into the locations of continents during the Paleozoic. When he plotted these locations, he found that glaciers of this time interval occurred in southern South America, southern Africa, southern India, Antarctica, and southern Australia. These places are now widely separated and, with the exception of Antarctica, do not currently lie in cold polar regions (**Fig. 2.2a**). To Wegener's amazement, all late Paleozoic glaciated areas lie *adjacent* to each other on his map of Pangaea. Furthermore, when he plotted the orientation of glacial striations, they all pointed roughly outward from a location in southeastern Africa. In other words, Wegener determined that the distribution of glaciations at the end of the Paleozoic Era could easily be explained if the continents had been united in Pangaea, with the southern part of Pangaea lying beneath the center of a huge ice cap. This distribution of glaciation could not be explained if the continents had always been in their present positions.

The Distribution of Climatic Belts

If the southern part of Pangaea had straddled the South Pole at the end of the Paleozoic Era, then during this same time interval, southern North America, southern Europe, and northwestern Africa would have straddled the equator and would have had tropical or subtropical climates. Wegener searched for evidence that this was so by studying sedimentary rocks that were formed at this time, for the material making up these rocks can reveal clues to the past climate. For example, in the swamps and jungles of tropical regions, thick deposits of plant material accumulate, and when deeply buried, this material transforms into coal. And in the clear, shallow seas of tropical regions, large reefs develop. Finally, subtropical regions, on either side of the tropical belt, contain deserts, an environment in which sand dunes form and salt from evaporating seawater or salt lakes accumulates. Wegener speculated that the distribution of late Paleozoic coal, reef, sand-dune, and salt deposits could define climate belts on Pangaea.

Sure enough, in the belt of Pangaea that Wegener expected to be equatorial, late Paleozoic sedimentary rock layers include abundant coal and the relics of reefs. And in the portions of Pangaea that Wegener predicted would be subtropical, late Paleozoic sedimentary rock layers include relics of desert dunes and deposits of salt (**Fig. 2.2b**). On a present-day map of our planet, exposures of these ancient rock layers scatter around the globe at a variety of latitudes. On Wegener's Pangaea, the exposures align in continuous bands that occupy appropriate latitudes.

The Distribution of Fossils

Today, different continents provide homes for different species. Kangaroos, for example, live only in Australia. Similarly, many kinds of plants grow only on one continent and not on others. Why? Because land-dwelling species of animals and plants cannot swim across vast oceans, and thus evolved independently on different continents. During a period of Earth history when all continents were in contact, however, land animals and plants could have migrated among many continents.

With this concept in mind, Wegener plotted fossil occurrences of land-dwelling species that existed during the late Paleozoic and early Mesozoic Eras (between about 300 and 210 million years ago) and found that these species had indeed existed on several continents (**Fig. 2.2c**). Wegener argued that the distribution of fossil species required the continents to have been adjacent to one another in the late Paleozoic and early Mesozoic Eras.

Matching Geologic Units

An art historian can recognize a Picasso painting, an architect knows what makes a building look "Victorian," and a geoscientist can identify a distinctive assemblage of rocks. Wegener found that the same distinctive Precambrian rock assemblages occurred on the eastern coast of South America and the western coast of Africa, regions now separated by an ocean (**Fig. 2.3a**). If the continents had been joined to create Pangaea in the past, then these matching rock groups would have been adjacent to each other, and thus could have composed

FIGURE 2.2 Wegener's evidence for continental drift came from analyzing the geologic record.

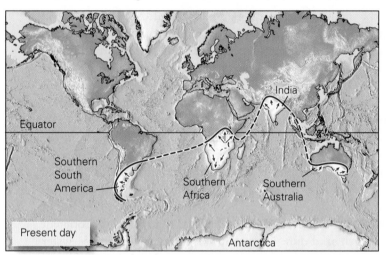

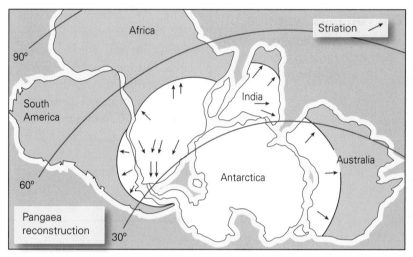

(a) The distribution of late Paleozoic glacial deposits and striations on present-day Earth are hard to explain. But on Pangaea, areas with glacial deposits fit together in a southern polar cap.

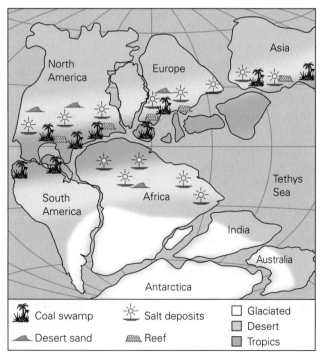

(b) The distribution of late Paleozoic rock types plots sensibly in the climate belts of Pangaea.

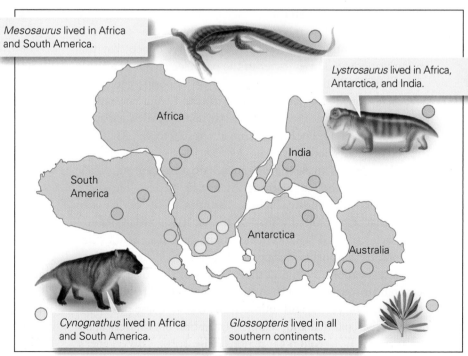

(c) A plot of fossil localities shows that Mesozoic land-dwelling organisms occur on multiple continents. This would be hard to explain if continents were separated.

continuous blocks or belts. Wegener also noted that features of the Appalachian Mountains of the United States and Canada closely resemble mountain belts in southern Greenland, Great Britain, Scandinavia, and northwestern Africa (**Fig. 2.3b, c**), regions that would have lain adjacent to each other in Pangaea. Wegener thus demonstrated that not only did the coastlines of continents match, so too did the rocks adjacent to the coastlines.

Criticism of Wegener's Ideas

Wegener's model of a supercontinent that later broke apart explained the distribution of ancient glaciers, coal, sand dunes, rock assemblages, and fossils. Clearly, he had compiled a strong *circumstantial* case for continental drift. But as noted earlier, he could not adequately explain how or why continents drifted. He left on his final expedition to Greenland having failed to convince his peers, and he died without knowing that his ideas, after lying dormant for decades, would be reborn as the basis of the broader theory of plate tectonics.

In effect, Wegener was ahead of his time. It would take three more decades of research before geologists obtained sufficient data to test his hypotheses properly. Collecting this data required instruments and techniques that did not exist in Wegener's day. Of the many geologic discoveries that ultimately opened the door to plate tectonics, perhaps the most important came from the discovery of a phenomenon called paleomagnetism, so we discuss it next.

> ### Take-Home Message
> Wegener argued that the continents were once merged into a supercontinent called Pangaea that later broke up to produce smaller continents that "drifted" apart. The matching shapes of coastlines, as well as the distribution of ancient climate belts, fossils, and rock units all make better sense if Pangaea existed.

2.3 Paleomagnetism and the Proof of Continental Drift

More than 1,500 years ago, Chinese sailors discovered that a piece of lodestone, when suspended from a thread, points in a northerly direction and can help guide a voyage. Lodestone exhibits this behavior because it consists of magnetite, an iron-rich mineral that, like a compass needle, aligns with Earth's magnetic field lines. While not as magnetic as lodestone, several other rock types contain tiny crystals of magnetite, or other magnetic minerals, and thus behave overall like weak magnets. In this section, we explain how the study of such magnetic behavior led to the realization that rocks preserve paleomagnetism, a record of Earth's magnetic field in the past. An understanding of paleomagnetism provided proof of continental drift and, as we'll see later in this chapter, contributed

FIGURE 2.3 Further evidence of drift: rocks on different sides of the ocean match.

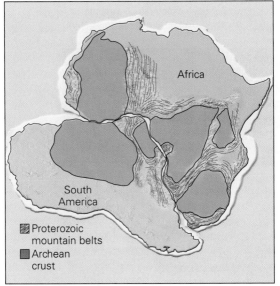

(a) Distinctive belts of rock in South America would align with similar ones in Africa, without the Atlantic.

Proterozoic mountain belts
Archean crust

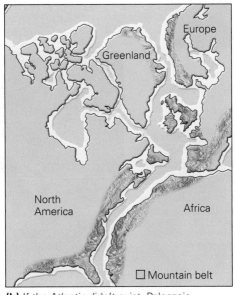

(b) If the Atlantic didn't exist, Paleozoic mountain belts on both coasts would be adjacent.

Mountain belt

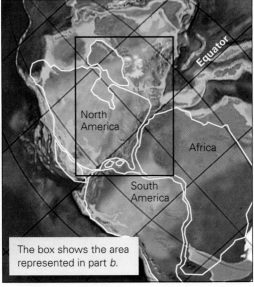

The box shows the area represented in part *b*.

(c) A modern reconstruction showing the positions of mountain belts in Pangaea. Modern continents are outlined in white.

to the development of plate tectonics theory. As a foundation for introducing paleomagnetism, we first provide additional detail about the basic nature of the Earth's magnetic field.

Earth's Magnetic Field

As we mentioned in Chapter 1, circulation of liquid iron alloy in the outer core of the Earth generates a magnetic field. (A similar phenomenon happens in an electrical dynamo at a power plant.) Earth's magnetic field resembles the field produced by a bar magnet, in that it has two ends of opposite polarity. Thus, we can represent Earth's field by a **magnetic dipole**, an imaginary arrow (**Fig. 2.4a**). Earth's dipole intersects the surface of the planet at two points, known as the **magnetic poles**. By convention, the

> **Did you ever wonder . . .**
> why compasses always point to the north?

north *magnetic* pole is at the end of the Earth nearest the north *geographic* pole (the point where the northern end of the spin axis intersects the surface). The north-seeking (red) end of a compass needle points to the north magnetic pole.

Earth's magnetic poles move constantly, but don't seem to stray further than about 1,500 km from the geographic poles, and averaged over thousands of years, they roughly coincide with Earth's geographic poles (**Fig. 2.4b**). That's because the rotation of the Earth causes the flow to organize into patterns resembling spring-like spirals, and these are roughly aligned with the spin axis. At present, the magnetic poles lie hundreds of kilometers away from the geographic poles, so the magnetic dipole tilts at about 11° relative to the Earth's spin axis. Because of this difference, a compass today does not point exactly to geographic north. The angle between the direction that a compass needle points and a line of longitude at a given location is the **magnetic declination** (**Fig. 2.4c**).

FIGURE 2.4 Features of Earth's magnetic field.

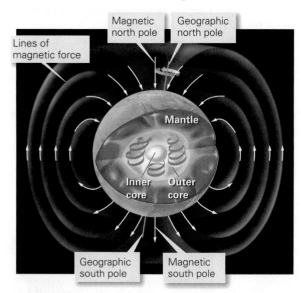

(a) The magnetic axis is not parallel to the spin axis. In 3-D, the Earth's magnetic field can be visualized as invisible curtains of energy, generated by flow to the outer core.

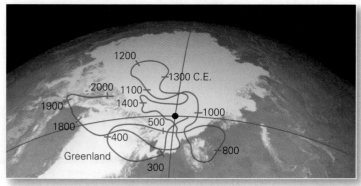

(b) A map of the magnetic pole position during the past 1,800 years shows that the pole moves, but stays within high latitudes.

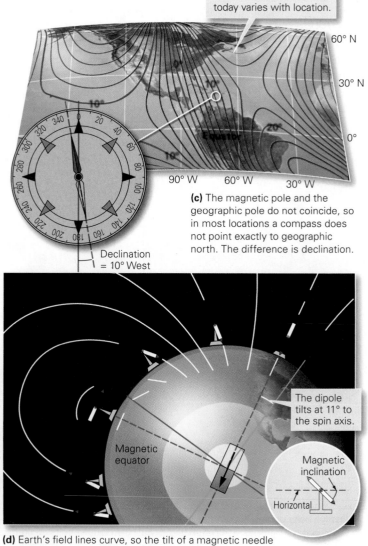

(c) The magnetic pole and the geographic pole do not coincide, so in most locations a compass does not point exactly to geographic north. The difference is declination.

The declination observed today varies with location.

Declination = 10° West

The dipole tilts at 11° to the spin axis.

Magnetic equator

Magnetic inclination

Horizontal

(d) Earth's field lines curve, so the tilt of a magnetic needle changes with latitude. This tilt is the magnetic inclination.

Invisible field lines curve through space between the magnetic poles. In a cross-sectional view, these lines lie parallel to the surface of the Earth (that is, are horizontal) at the equator, tilt at an angle to the surface in midlatitudes, and plunge perpendicular to the surface at the magnetic poles (**Fig. 2.4d**). The angle between a magnetic field line and the surface of the Earth, at a given location, is called the **magnetic inclination**. If you place a magnetic needle on a horizontal axis so that it can pivot up and down, and then carry it from the magnetic equator to the magnetic pole, you'll see that the inclination varies with latitude—it is 0° at the magnetic equator and 90° at the magnetic poles. (Note that the compass you may carry with you on a hike does not show inclination because it has been balanced to remain horizontal.)

What Is Paleomagnetism?

In the early 20th century, researchers developed instruments that could measure the weak magnetic field produced by rocks and made a surprising discovery. In a rock that formed millions of years ago, the orientation of the dipole representing the magnetic field of the rock *is not the same* as that of present-day Earth (**Fig. 2.5a**). To understand this statement, consider an example. Imagine traveling to a location near the coast on the equator in South America where the inclination and declination are presently 0°. If you measure the weak magnetic field produced by, say, a 90-million-year-old rock, and represent the

orientation of this field by an imaginary bar magnet, you'll find that this imaginary bar magnet does *not* point to the present-day north magnetic pole, and you'll find that its inclination is not 0°. The reason for this difference is that the magnetic fields of ancient rocks indicate the orientation of the magnetic field, relative to the rock, at the time the rock formed. This record, preserved in rock, is **paleomagnetism**.

Paleomagnetism can develop in many different ways. For example, when lava, molten rock containing no crystals, starts to cool and solidify into rock, tiny magnetite crystals begin to grow (**Fig. 2.5b**). At first, thermal energy causes the tiny magnetic dipole associated with each crystal to wobble and tumble chaotically. Thus, at any given instant, the dipoles of the magnetite specks are randomly oriented and the magnetic forces they produce cancel each other out. Eventually, however, the rock cools sufficiently that the dipoles slow down and, like tiny compass needles, align with the Earth's magnetic field. As the rock cools still more, these tiny compass needles lock into permanent parallelism with the Earth's magnetic field at the time the cooling takes place. Since the magnetic dipoles of all the grains point in the same direction, they add together and produce a measurable field.

Apparent Polar Wander— A Proof That Continents Move

Why doesn't the paleomagnetic dipole in ancient rocks point to the present-day magnetic field? When geologists first

FIGURE 2.5 Paleomagnetism and how it can form during the solidification and cooling of lava.

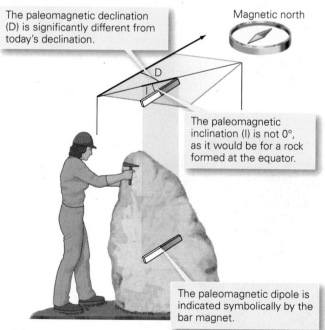

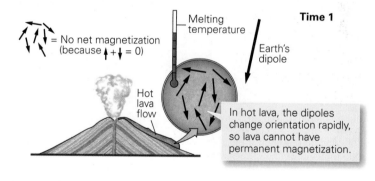

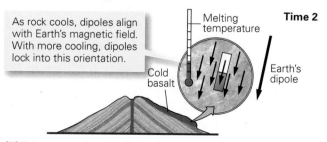

(a) A geologist finds an ancient rock sample at a location on the equator, where declination today is 0°. The orientation of the rock's paleomagnetism is different from that of today's field.

(b) Paleomagnetism can form when lava cools and becomes solid rock.

attempted to answer this question, they *assumed* that continents were fixed in position and thus concluded that the positions of Earth's magnetic poles in the past were different than they are today. They introduced the term **paleopole** to refer to the supposed position of the Earth's magnetic north pole in the past. With this concept in mind, they set out to track what they thought was the change in position of the paleopole over time. To do this, they measured the paleomagnetism in a succession of rocks of different ages from the same general location on a continent, and they plotted the position of the associated succession of paleopole positions on a map (**Fig. 2.6a**). The successive positions of dated paleopoles trace out a curving line that came to be known as an **apparent polar-wander path.**

At first, geologists assumed that the apparent polar-wander path actually represented how the position of Earth's magnetic pole migrated through time. But were they in for a surprise! When they obtained polar-wander paths from many different continents, they found that *each continent has a different apparent polar-wander path*. The hypothesis that continents are fixed in position cannot explain this observation, for if the

magnetic pole moved while all the continents stayed fixed, measurements from all continents should produce the same apparent polar-wander paths.

Geologists suddenly realized that they were looking at apparent polar-wander paths in the wrong way. It's not the pole that moves relative to fixed continents, but rather the continents that move relative to a fixed pole (**Fig. 2.6b**). Since each continent has its own unique polar-wander path (**Fig. 2.6c**), *the continents must move with respect to each other.* The discovery proved that Wegener was essentially right all along—continents do move!

Take-Home Message

Study of paleomagnetism indicates that the continents have moved relative to the Earth's magnetic poles. Each continent has a different apparent polar-wander path, which is only possible if the continents move ("drift") relative to each other.

FIGURE 2.6 Apparent polar-wander paths and their interpretation.

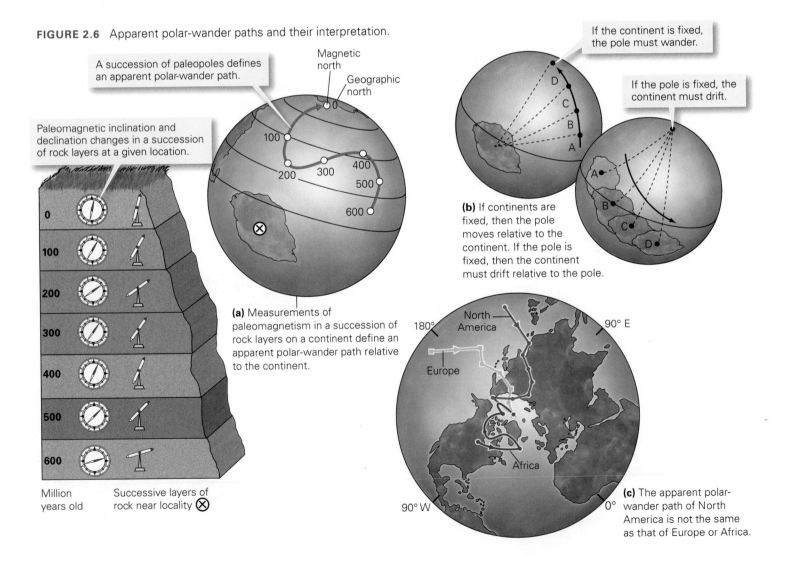

A succession of paleopoles defines an apparent polar-wander path.

Paleomagnetic inclination and declination changes in a succession of rock layers at a given location.

(a) Measurements of paleomagnetism in a succession of rock layers on a continent define an apparent polar-wander path relative to the continent.

Million years old / Successive layers of rock near locality ⊗

If the continent is fixed, the pole must wander.

If the pole is fixed, the continent must drift.

(b) If continents are fixed, then the pole moves relative to the continent. If the pole is fixed, then the continent must drift relative to the pole.

(c) The apparent polar-wander path of North America is not the same as that of Europe or Africa.

2.4 The Discovery of Sea-Floor Spreading

New Images of Sea-Floor Bathymetry

Military needs during World War II gave a boost to sea-floor exploration, for as submarine fleets grew, navies required detailed information about **bathymetry**, or depth variations. The invention of echo sounding (sonar) permitted such information to be gathered quickly. Echo sounding works on the same principle that a bat uses to navigate and find insects. A sound pulse emitted from a ship travels down through the water, bounces off the sea floor, and returns up as an echo through the water to a receiver on the ship. Since sound waves travel at a known velocity, the time between the sound emission and the echo detection indicates the distance between the ship and the sea floor. (Recall that velocity = distance/time, so distance = velocity × time.) As the ship travels, observers can obtain a continuous record of the depth of the sea floor. The resulting cross section showing depth plotted against location is called a bathymetric profile (**Fig. 2.7a, b**). By cruising back and forth across the ocean many times, investigators obtained a series of bathymetric profiles and from these constructed maps of the sea floor. (Geologists can now produce such maps much more rapidly using satellite data.) Bathymetric maps reveal several important features.

> *Mid-ocean ridges*: The floor beneath all major oceans includes **abyssal plains**, which are broad, relatively flat regions of the ocean that lie at a depth of about 4 to 5 km below sea level; and **mid-ocean ridges**, submarine mountain ranges whose peaks lie only about 2 to 2.5 km below sea level (**Fig. 2.8a**). Geologists call the crest of the mid-ocean ridge the ridge axis. All mid-ocean ridges are roughly symmetrical—bathymetry on one side of the axis is nearly a mirror image of bathymetry on the other side.

> *Deep-ocean trenches*: Along much of the perimeter of the Pacific Ocean, and in a few other localities as well, the ocean floor reaches depths of 8 to 12 km—deep enough to swallow Mt. Everest. These deep areas occur in elongate troughs that are now referred to as **trenches** (**Fig. 2.8b**). Trenches border **volcanic arcs**, curving chains of active volcanoes.

> *Seamount chains*: Numerous volcanic islands poke up from the ocean floor: for example, the Hawaiian Islands lie in the middle of the Pacific. In addition to islands that rise above sea level, sonar has detected many **seamounts** (isolated submarine mountains), which were once volcanoes but no longer erupt. Volcanic islands and seamounts typically occur in chains, but in contrast to the volcanic arcs that border deep-ocean trenches, only one island at the end of a seamount and island chain remains capable of erupting volcanically today.

> *Fracture zones*: Surveys reveal that the ocean floor is diced up by narrow bands of vertical cracks and broken-up rock. These **fracture zones** lie roughly at right angles to mid-ocean ridges. The ridge axis typically steps sideways when it intersects with a fracture zone.

FIGURE 2.7 Bathymetry of mid-ocean ridges and abyssal plains.

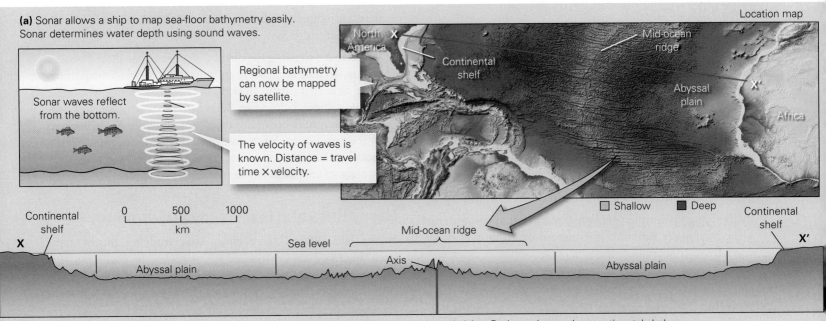

(a) Sonar allows a ship to map sea-floor bathymetry easily. Sonar determines water depth using sound waves.

Location map

Sonar waves reflect from the bottom.

Regional bathymetry can now be mapped by satellite.

The velocity of waves is known. Distance = travel time × velocity.

North America · Continental shelf · Mid-ocean ridge · Abyssal plain · Africa

☐ Shallow ■ Deep

Continental shelf · Abyssal plain · Sea level · Mid-ocean ridge · Axis · Abyssal plain · Continental shelf

0 500 1000
km

(b) A bathymetric profile along line X–X′ illustrates how mid-ocean ridges rise above abyssal plains. Both are deeper than continental shelves.

FIGURE 2.8 Other bathymetric features of the ocean floor.

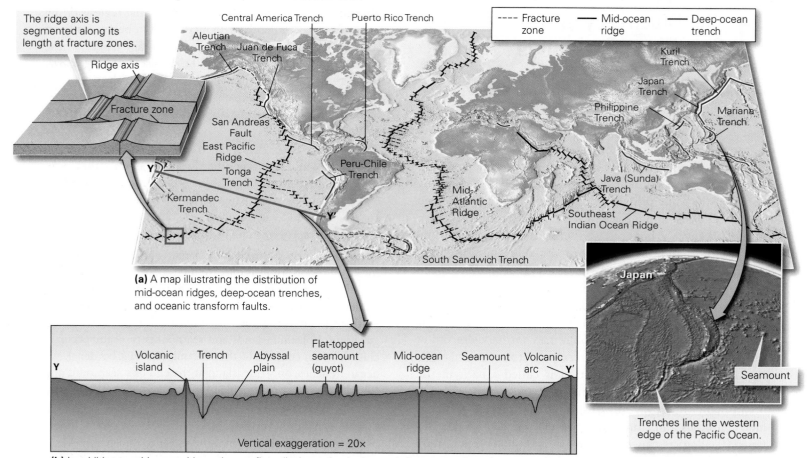

The ridge axis is segmented along its length at fracture zones.

Ridge axis

Fracture zone

Central America Trench

Puerto Rico Trench

Aleutian Trench

Juan de Fuca Trench

San Andreas Fault

East Pacific Ridge

Tonga Trench

Kermadec Trench

Peru-Chile Trench

Mid-Atlantic Ridge

Kuril Trench

Japan Trench

Philippine Trench

Mariana Trench

Java (Sunda) Trench

Southeast Indian Ocean Ridge

South Sandwich Trench

---- Fracture zone   —— Mid-ocean ridge   —— Deep-ocean trench

(a) A map illustrating the distribution of mid-ocean ridges, deep-ocean trenches, and oceanic transform faults.

Japan

Seamount

Trenches line the western edge of the Pacific Ocean.

Y   Volcanic island   Trench   Abyssal plain   Flat-topped seamount (guyot)   Mid-ocean ridge   Seamount   Volcanic arc   Y'

Vertical exaggeration = 20×

(b) In addition to mid-ocean ridges, the sea floor displays other bathymetric features such as deep-sea trenches, oceanic islands, guyots, and seamounts, as shown in this vertically exaggerated profile.

New Observations on the Nature of Oceanic Crust

By the mid-20th century, geologists had discovered many important characteristics of the sea-floor crust. These discoveries led them to realize that oceanic crust differs from continental crust, and that bathymetric features of the ocean floor provide clues to the origin of the crust. Specifically:

> A layer of sediment composed of clay and the tiny shells of dead plankton covers much of the ocean floor. This layer becomes progressively thicker away from the mid-ocean ridge axis. But even at its thickest, the sediment layer is too thin to have been accumulating for the entirety of Earth history.

> By dredging up samples, geologists learned that oceanic crust is fundamentally different in composition from continental crust. Beneath its sediment cover, oceanic crust bedrock consists primarily of basalt—it does not display the great variety of rock types found on continents.

> Heat flow, the rate at which heat rises from the Earth's interior up through the crust, is not the same everywhere in the oceans. Rather, more heat rises beneath mid-ocean ridges than elsewhere. This observation led researchers to speculate that hot magma might be rising into the crust just below the mid-ocean ridge axis.

> When maps showing the distribution of earthquakes in oceanic regions became available in the years after World War II, it became clear that earthquakes do not occur randomly, but rather define distinct belts (**Fig. 2.9**). Some belts follow trenches, some follow mid-ocean ridge axes, and others lie along portions of fracture zones. Since earthquakes define locations where rocks break and move, geologists realized that these bathymetric features are places where motion is taking place.

Harry Hess and His "Essay in Geopoetry"

In the late 1950s, Harry Hess, after studying the observations described above, realized that because the sediment layer on the ocean floor was thin overall, the ocean floor might be much younger than the continents. Also, because the sediment thickened progressively away from mid-ocean ridges, the ridges themselves likely were younger than the deeper parts of the

FIGURE 2.9 A 1953 map showing the distribution of earthquake locations in the ocean basins. Note that earthquakes occur in belts.

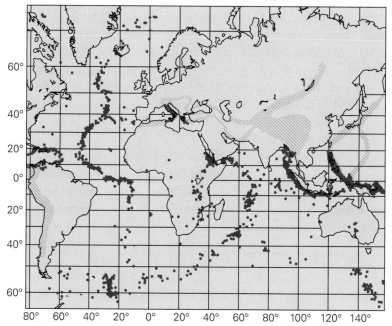

into the mantle. Hess suggested that earthquakes at trenches were evidence of this movement, but he didn't understand how the movement took place. Other geologists, such as Robert Dietz, were coming to similar conclusions at about the same time.

Hess and his contemporaries realized that the sea-floor-spreading hypothesis instantly provided the long-sought explanation of how continental "drift" occurs. Continents passively move apart as the sea floor between them spreads at mid-ocean ridges, and they passively move together as the sea floor between them sinks back into the mantle at trenches. (As we will see later, geologists now realize that it is the lithosphere that moves, not just the crust.) Thus, sea-floor spreading proved to be an important step on the route to plate tectonics—the idea seemed so good that Hess referred to his description of it as "an essay in geopoetry." But first, the idea needed to be tested, and other key discoveries would have to take place before the whole theory of plate tectonics could come together.

> **Did you ever wonder...**
> if the distance between New York and Paris changes?

ocean floor. If this was so, then somehow *new ocean floor must be forming at the ridges*, and thus an ocean basin could be getting wider with time. But how? The association of earthquakes with mid-ocean ridges suggested to him that the sea floor was cracking and splitting apart at the ridge. The discovery of high heat flow along mid-ocean ridge axes provided the final piece of the puzzle, for it suggested the presence of very hot molten rock beneath the ridges. In 1960, Hess suggested that indeed molten rock (basaltic magma) rose upward beneath mid-ocean ridges and that this material solidified to form oceanic crust basalt (**Fig. 2.10**). The new sea floor then moved away from the ridge, a process we now call **sea-floor spreading**. Hess realized that old ocean floor must be consumed somewhere, or the Earth would have to be expanding, so he suggested that deep-ocean trenches might be places where the sea floor sank back

> ## Take-Home **Message**
> New studies of the sea floor led to the proposal of sea-floor spreading. New sea floor forms at mid-ocean ridges and then moves away from the axis, so ocean basins can get wider with time. Old ocean floor sinks back into the mantle by subduction. As ocean basins grow or shrink, continents drift.

2.5 Evidence for Sea-Floor Spreading

For a hypothesis to become a theory (see Box P.1), researchers must demonstrate that the idea really works. During the 1960s, geologists found that the sea-floor spreading hypothesis successfully explains several previously baffling observations. Here we discuss two: (1) the existence of orderly variations in the strength of the measured magnetic field over the sea floor, producing a pattern of stripes called marine magnetic anomalies; and (2) the variation in sediment thickness on the ocean crust, as measured by drilling.

FIGURE 2.10 Harry Hess's basic concept of sea-floor spreading. Hess implied, incorrectly, that only the crust moved. We will see that this sketch is an oversimplification.

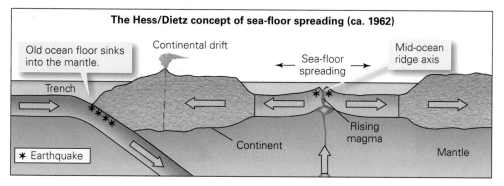

Marine Magnetic Anomalies

Recognizing anomalies. Geologists can measure the strength of Earth's magnetic field with an instrument called a magnetometer. At any given location on the surface of the Earth, the magnetic field that you measure includes two parts: one produced by the main dipole of the Earth generated by circulation of molten iron in the outer core, and another produced by the magnetism of near-surface rock. A **magnetic anomaly** is the difference between the *expected* strength of the Earth's main dipole field at a certain location and the *actual* measured strength of the magnetic field at that location. Places where the field strength is stronger than expected are *positive* anomalies, and places where the field strength is weaker than expected are *negative* anomalies.

Geologists towed magnetometers back and forth across the ocean to map variations in magnetic field strength (**Fig. 2.11a**). As a ship cruised along its course, the magnetometer's gauge might first detect an interval of strong signal (a positive anomaly) and then an interval of weak signal (a negative anomaly). A graph of signal strength versus distance along the traverse, therefore, has a sawtooth shape (**Fig. 2.11b**). When geologists compiled data from many cruises on a map, these **marine magnetic anomalies** defined distinctive, alternating bands. If we color positive anomalies dark and negative anomalies light, the pattern made by the anomalies resembles the stripes on a candy cane (**Fig. 2.11c**). The mystery of this marine magnetic anomaly pattern, however, remained unsolved until geologists recognized the existence of magnetic reversals.

Magnetic reversals. Recall that Earth's magnetic field can be represented by an arrow, representing the dipole, that presently points from the north magnetic pole to the south magnetic pole. When researchers measured the paleomagnetism of a succession of rock layers that had accumulated over a long period of time, they found that the polarity (which end of a magnet points north and which end points south) of the paleomagnetic field preserved in some layers was the same as that of Earth's present magnetic field, whereas in other layers it was the opposite (**Fig. 2.12a, b**).

At first, observations of reversed polarity were largely ignored, thought to be the result of lightning strikes or of local chemical reactions between rock and water. But when repeated measurements from around the world revealed a systematic pattern of alternating normal and reversed polarity in rock layers, geologists realized that reversals were a worldwide, not a local, phenomenon. They reached the unavoidable conclusion that, at various times during Earth history, *the polarity of Earth's magnetic field has suddenly reversed!* In other words, sometimes the Earth has normal polarity, as it does today, and sometimes it has reversed polarity (**Fig. 2.12c**). A time when the Earth's field flips from normal to reversed polarity, or vice versa, is called a **magnetic reversal**. When the Earth has reversed polarity, the south magnetic pole lies near the north geographic pole, and the north magnetic pole lies near the south geographic pole. Thus, if you were to use a compass during periods when the Earth's magnetic field was reversed, the north-seeking end of the needle would point to the south

FIGURE 2.11 The discovery of marine magnetic anomalies.

(a) A ship towing a magnetometer detects changes in the strength of the magnetic field.

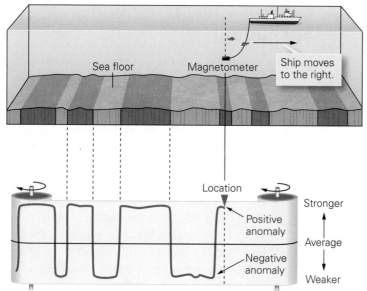

(b) On a paper record, intervals of stronger magnetism (positive anomalies) alternate with intervals of weaker magnetism (negative anomalies).

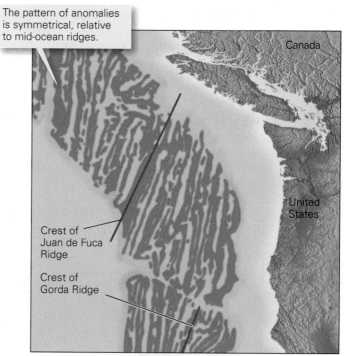

(c) A map showing areas of positive anomalies (dark) and negative anomalies (light) off the west coast of North America. The pattern of anomalies resembles candy-cane strips.

geographic pole. Note that the Earth itself doesn't turn upside down—it is just the magnetic field that reverses.

In the 1950s, about the same time researchers discovered polarity reversals, they developed a technique that permitted them to measure the age of a rock in years. The technique, called iso-topic dating, will be discussed in detail in Chapter 10. Geologists applied the technique to determine the ages of rock layers in which they obtained their paleomagnetic measurements, and thus determined *when* the magnetic field of the Earth reversed. With this information, they constructed a history of magnetic reversals for the past 4.5 million years; this history is now called the magnetic-reversal chronology. The time interval between successive reversals is called a **chron**.

A diagram representing the Earth's magnetic-reversal chronology (**Fig. 2.12d**) shows that reversals do not occur regu-larly, so the lengths of different polarity chrons are different. For example, we have had a normal-polarity chron for about the last 700,000 years. Before that, a reversed-polarity chron occurred. The youngest four polarity chrons (Brunhes, Matuyama, Gauss, and Gilbert) were named after scientists who had made

FIGURE 2.12 Magnetic polarity reversals, and the chronology of reversals.

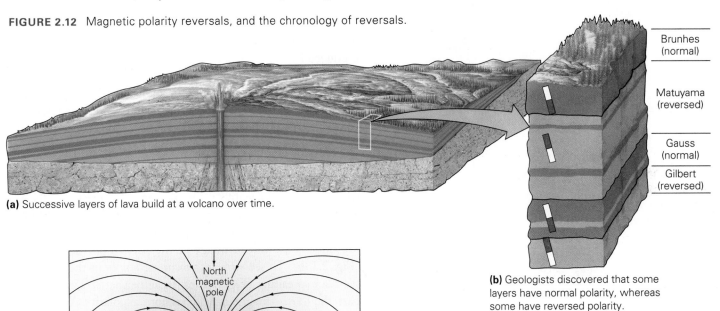

(a) Successive layers of lava build at a volcano over time.

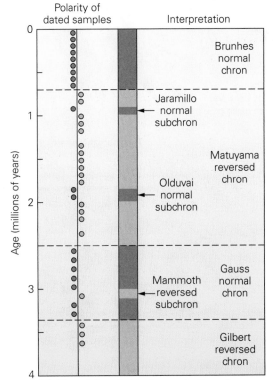

(b) Geologists discovered that some layers have normal polarity, whereas some have reversed polarity.

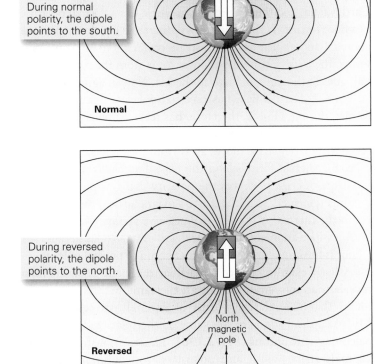

(c) Geologists proposed that the Earth's magnetic field reverses polarity every now and then.

(d) Observations led to the production of a reversal chronology, with named polarity intervals.

FIGURE 2.13 The progressive development of magnetic anomalies, and the long-term reversal chronology.

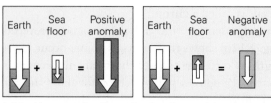

(a) Positive anomalies form when sea-floor rock has the same polarity as the present magnetic field. Negative anomalies form when sea-floor rock has polarity that is opposite to the present field.

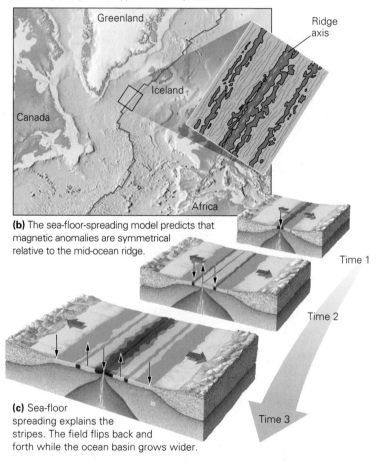

(b) The sea-floor-spreading model predicts that magnetic anomalies are symmetrical relative to the mid-ocean ridge.

(c) Sea-floor spreading explains the stripes. The field flips back and forth while the ocean basin grows wider.

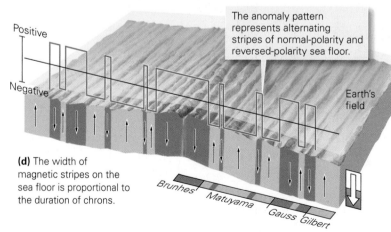

The anomaly pattern represents alternating stripes of normal-polarity and reversed-polarity sea floor.

(d) The width of magnetic stripes on the sea floor is proportional to the duration of chrons.

important contributions to the study of magnetism. As more measurements became available, investigators realized that some short-duration reversals (less than 200,000 years long) took place within the chrons, and they called these shorter durations "polarity subchrons." Using isotopic dating, it was possible to determine the age of chrons back to 4.5 Ma.

Interpreting marine magnetic anomalies.   Why do marine magnetic anomalies exist? In 1963, researchers in Britain and Canada proposed a solution to this riddle. Simply put, a positive anomaly occurs over areas of the sea floor where underlying basalt has normal polarity. In these areas, the magnetic force produced by the magnetite grains in basalt *adds* to the force produced by the Earth's dipole—the sum of these forces yields a stronger magnetic signal than expected due to the dipole alone (**Fig. 2.13a**). A negative anomaly occurs over regions of the sea floor where the underlying basalt has a reversed polarity. In these regions, the magnetic force of the basalt *subtracts* from the force produced by the Earth's dipole, so the measured magnetic signal is weaker than expected.

The sea-floor-spreading model easily explains not only why positive and negative magnetic anomalies exist over the sea floor, but also why they define stripes that trend parallel to the mid-ocean ridge and why the pattern of stripes on one side of the ridge is the mirror image of the pattern on the other side (**Fig. 2.13b**). To see why, let's examine stages in the process of sea-floor spreading (**Fig. 2.13c**). Imagine that at Time 1 in the past, the Earth's magnetic field has normal polarity. As the basalt rising at the mid-ocean ridge during this time interval cools and solidifies, the tiny magnetic grains in basalt align with the Earth's field, and thus the rock as a whole has a normal polarity. Sea floor formed during Time 1 will therefore generate a positive anomaly and appear as a dark stripe on an anomaly map. As it forms, the rock of this stripe moves away from the ridge axis, so half goes to the right and half to the left. Now imagine that later, at Time 2, Earth's field has reversed polarity. Sea-floor basalt formed during Time 2, therefore, has reversed polarity and will appear as a light stripe on an anomaly map. As it forms, this reversed-polarity stripe moves away from the ridge axis, and even younger crust forms along the axis. The basalt in each new stripe of crust preserves the polarity that was present at the time it formed, so as the Earth's magnetic field flips back and forth, alternating positive and negative anomaly stripes form. A positive anomaly exists over the ridge axis today because sea floor is forming during the present chron of normal polarity.

Closer examination of a sea-floor magnetic anomaly map reveals that anomalies are not all the same width. Geologists found that the relative widths of anomaly stripes near the Mid-Atlantic Ridge are the same as the relative durations of paleomagnetic chrons (**Fig. 2.13d**). This relationship between anomaly-stripe width and polarity-chron duration indicates that the rate of sea-floor spreading has been constant along the Mid-Atlantic Ridge for at least the last 4.5 million years. If you assume that the spreading rate was constant for tens to hundreds of millions of years, then it is possible to estimate the age of stripes right up to the edge of the ocean.

Evidence from Deep-Sea Drilling

In the late 1960s, a research drilling ship called the *Glomar Challenger* set out to sail around the ocean drilling holes into the sea floor. This amazing ship could lower enough drill pipe to drill in 5-km-deep water and could continue to drill until the hole reached a depth of about 1.7 km (1.1 miles) below the sea floor. Drillers brought up cores of rock and sediment that geoscientists then studied on board.

On one of its early cruises, the *Glomar Challenger* drilled a series of holes through sea-floor sediment to the basalt layer. These holes were spaced at progressively greater distances from the axis of the Mid-Atlantic Ridge. If the model of sea-floor spreading was correct, then not only should the sediment layer be progressively thicker away from the axis, but the age of the oldest sediment just above the basalt should be progressively older away from the axis. When the drilling and the analyses were complete, the prediction was confirmed. Thus, studies of both marine magnetic anomalies and the age of the sea floor proved the sea-floor-spreading model.

> ## Take-Home **Message**
> Marine magnetic anomalies form because reversals of the Earth's magnetic polarity take place while sea-floor spreading occurs. The discovery of these anomalies, as well as documentation that the sea floor gets older away from the ridge axis, proved that the sea-floor-spreading hypothesis is correct.

2.6 What Do We Mean by Plate Tectonics?

The paleomagnetic proof of continental drift and the discovery of sea-floor spreading set off a scientific revolution in geology in the 1960s and 1970s. Geologists realized that many of their existing interpretations of global geology, based on the premise that the positions of continents and oceans remain fixed in position through time, were simply wrong! Researchers dropped what they were doing and turned their attention to studying the broader implications of continental drift and sea-floor spreading. It became clear that these phenomena

FIGURE 2.14 Nature of the lithosphere and its behavior.

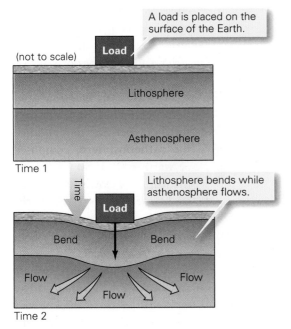

(a) The lithosphere is fairly rigid, but when a heavy load, such as a glacier or volcano, builds on its surface, the surface bends down. This can happen because the underlying "plastic" asthenosphere can flow out of the way.

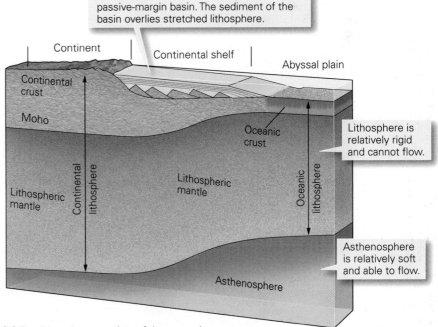

(b) The lithosphere consists of the crust plus the uppermost mantle. It is thicker beneath continents than beneath oceans.

FIGURE 2.15 The locations of plate boundaries and the distribution of earthquakes.

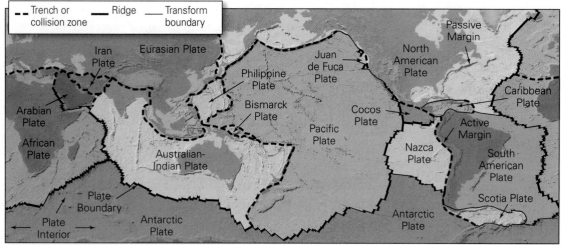

(a) A map of major plates shows that some consist entirely of oceanic lithosphere, whereas some consist of both continental and oceanic lithosphere. Active continental margins lie along plate boundaries; passive margins do not.

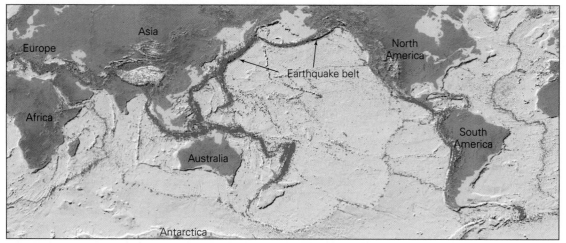

(b) The locations of earthquakes (red dots) mostly fall in distinct bands that correspond to plate boundaries. Relatively few earthquakes occur in the stabler plate interiors.

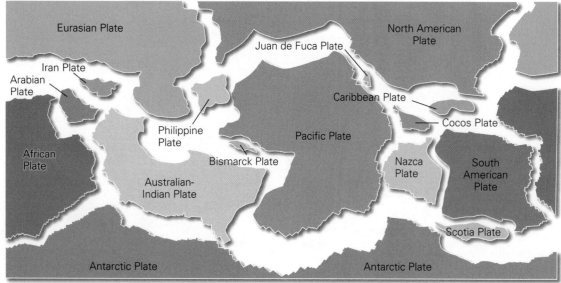

(c) An exploded view of the plates emphasizes the variation in shape and size of the plates.

required that the outer shell of the Earth was divided into rigid plates that moved relative to each other. New studies clarified the meaning of a plate, defined the types of plate boundaries, constrained plate motions, related plate motions to earthquakes and volcanoes, showed how plate interactions can explain mountain belts and seamount chains, and outlined the history of past plate motions. From these, the modern theory of plate tectonics evolved. Below, we first describe lithosphere plates and their boundaries, and then outline the basic principles of plate tectonics theory.

The Concept of a Lithosphere Plate

We learned earlier that geoscientists divide the outer part of the Earth into two layers. The **lithosphere** consists of the crust plus the top (cooler) part of the upper mantle. It behaves relatively rigidly, meaning that when a force pushes or pulls on it, it does not flow but rather bends or breaks (**Fig. 2.14a**). The lithosphere floats on a relatively soft, or "plastic," layer called the **asthenosphere**, composed of warmer (> 1280°C) mantle that can flow slowly when acted on by a force. As a result, the asthenosphere convects, like water in a pot, though much more slowly.

Continental lithosphere and oceanic lithosphere differ markedly in their thicknesses. On average, continental lithosphere has a thickness of 150 km, whereas old oceanic lithosphere has a thickness of about 100 km (**Fig. 2.14b**).

(For reasons discussed later in this chapter, new oceanic lithosphere at a mid-ocean ridge is much thinner.) Recall that the crustal part of continental lithosphere ranges from 25 to 70 km thick and consists largely of low-density felsic and intermediate rock (see Chapter 1). In contrast, the crustal part of oceanic lithosphere is only 7 to 10 km thick and consists largely of relatively high-density mafic rock (basalt and gabbro). The mantle part of both continental and oceanic lithosphere consists of very high-density ultramafic rock (peridotite). Because of these differences, the continental lithosphere "floats" at a higher level than does the oceanic lithosphere.

The lithosphere forms the Earth's relatively rigid shell. But unlike the shell of a hen's egg, the lithospheric shell contains a number of major breaks, which separate it into distinct pieces. As noted earlier, we call the pieces **lithosphere plates**, or simply plates. The breaks between plates are known as **plate boundaries** (**Fig. 2.15a**). Geoscientists distinguish twelve major plates and several microplates.

The Basic Principles of Plate Tectonics

With the background provided above, we can restate plate tectonics theory concisely as follows. The Earth's lithosphere is divided into plates that move relative to each other. As a plate moves, its internal area remains mostly, but not perfectly, rigid and intact. But rock along plate boundaries undergoes intense deformation (cracking, sliding, bending, stretching, and squashing) as the plate grinds or scrapes against its neighbors or pulls away from its neighbors. As plates move, so do the continents that form part of the plates. Because of plate tectonics, the map of Earth's surface constantly changes.

Identifying Plate Boundaries

How do we recognize the location of a plate boundary? The answer becomes clear from looking at a map showing the locations of earthquakes (**Fig. 2.15b**). Recall from Chapter 1 that earthquakes are vibrations caused by shock waves that are generated where rock breaks and suddenly slips along a fault. The epicenter marks the point on the Earth's surface directly above the earthquake. Earthquake epicenters do not speckle the globe randomly, like buckshot on a target. Rather, the majority occur in relatively narrow, distinct belts. These earthquake belts define the position of plate boundaries because the fracturing and slipping that occurs along plate boundaries generates earthquakes. Plate interiors, regions away from the plate boundaries, remain relatively earthquake-free because they do not accommodate as much movement. While earthquakes serve as the most definitive indicator of a plate boundary, other prominent geologic features also develop along plate boundaries, as you will learn by the end of this chapter.

Note that some plates consist entirely of oceanic lithosphere, whereas some plates consist of both oceanic and continental lithosphere. Also, note that not all plates are the same size (**Fig. 2.15c**). Some plate boundaries follow continental margins, the boundary between a continent and an ocean, but others do not. For this reason, we distinguish between **active margins**, which are plate boundaries, and **passive margins**, which are not plate boundaries. Earthquakes are common at active margins, but not at passive margins. Along passive margins, continental crust is thinner than in continental interiors. Thick (10 to 15 km) accumulations of sediment cover this thinned crust. The surface of this sediment

> **Did you ever wonder...**
> why earthquakes don't occur everywhere?

FIGURE 2.16 The three types of plate boundaries differ based on the nature of relative movement.

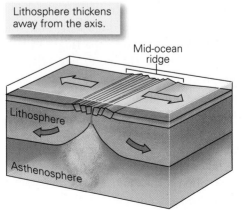

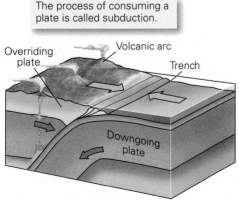

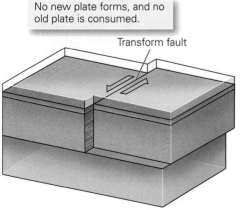

(a) At a divergent boundary, two plates move away from the axis of a mid-ocean ridge. New oceanic lithosphere forms.

(b) At a convergent boundary, two plates move toward each other; the downgoing plate sinks beneath the overriding plate.

(c) At a transform boundary, two plates slide past each other on a vertical fault surface.

layer is a broad, shallow (less than 500 m deep) region called the **continental shelf**, home to the major fisheries of the world.

Geologists define three types of plate boundaries, based simply on the relative motions of the plates on either side of the boundary (**Fig. 2.16a–c**). A boundary at which two plates move apart from each other is a **divergent boundary**. A boundary at which two plates move toward each other so that one plate sinks beneath the other is a **convergent boundary**. And a boundary at which two plates slide sideways past each other is a **transform boundary**. Each type of boundary looks and behaves differently from the others, as we will now see.

> ## Take-Home Message
>
> Earth's lithosphere is divided into about twenty plates that move relative to each other. Geologists recognize three different types of plate boundaries (divergent, convergent, and transform) based on relative motion across the boundary. Plate boundaries are defined by seismic belts.

2.7 Divergent Plate Boundaries and Sea-Floor Spreading

At a divergent boundary, or spreading boundary, two oceanic plates move apart by the process of sea-floor spreading. Note that an *open space does not develop* between diverging plates. Rather, as the plates move apart, new oceanic lithosphere forms continually along the divergent boundary (**Fig. 2.17a**). This process takes place at a submarine mountain range called a mid-ocean ridge that rises 2 km above the adjacent abyssal plains of the ocean. Thus, geologists commonly refer to a divergent boundary as a mid-ocean ridge, or simply a ridge. Water depth above ridges averages about 2.5 km.

To characterize a divergent boundary more completely, let's look at one mid-ocean ridge in more detail (**Fig. 2.17b**). The Mid-Atlantic Ridge extends from the waters between northern Greenland and northern Scandinavia southward across the equator to the latitude of the southern tip of South America. Geologists have found that the formation of new sea floor takes place *only* along the axis (centerline) of the ridge, which is marked by an elongate valley. The sea floor slopes away, reaching the depth of the abyssal plain (4 to 5 km) at a distance of about 500 to 800 km from the ridge axis (see Fig. 2.7). Roughly speaking, the Mid-Atlantic Ridge is symmetrical—its eastern half looks like a mirror image of its western half. The ridge consists, along its length, of short segments (tens to hundreds of km long) that step over at breaks that, as we noted earlier, are called fracture zones. Later, we will see that these correspond to transform faults.

How Does Oceanic Crust Form at a Mid-Ocean Ridge?

As sea-floor spreading takes place, hot asthenosphere rises beneath the ridge and begins to melt, and molten rock, or magma, forms (**Fig. 2.17c**). We will explain why this magma forms in Chapter 4. Magma has a lower density than solid rock, so it behaves buoyantly and rises, as oil rises above vinegar in salad dressing. Molten rock eventually accumulates in the crust below the ridge axis, filling a region called a magma chamber. As the magma cools, it turns into a mush of crystals. Some of the magma solidifies completely along the side of the chamber to make the coarse-grained, mafic igneous rock called gabbro. The rest rises still higher to fill vertical cracks, where it solidifies and forms wall-like sheets, or dikes, of basalt. Some magma rises all the way to the surface of the sea floor at the ridge axis and spills out of small submarine volcanoes. The resulting lava cools to form a layer of basalt blobs called pillows. Observers in research submarines have detected chimneys spewing hot, mineralized water rising from cracks in the sea floor along the ridge axis. These chimneys are called **black smokers** because the water they emit looks like a cloud of dark smoke; the color comes from a suspension of tiny mineral grains that precipitate in the water the instant that the water cools (**Fig. 2.18**).

As soon as it forms, new oceanic crust moves away from the ridge axis, and when this happens, more magma rises from below, so still more crust forms. In other words, like a vast, continuously moving conveyor belt, magma from the mantle rises to the Earth's surface at the ridge, solidifies to form oceanic crust, and then moves laterally away from the ridge. Because all sea floor forms at mid-ocean ridges, the youngest sea floor occurs on either side of the ridge axis, and sea floor becomes progressively older away from the ridge. In the Atlantic Ocean, the oldest sea floor, therefore, lies adjacent to the passive continental margins on either side of the ocean (**Fig. 2.19**). The oldest ocean floor on our planet underlies the western Pacific Ocean; this crust formed about 200 million years ago.

The tension (stretching force) applied to newly formed solid crust as spreading takes place breaks the crust, resulting in the formation of faults. Slip on the faults causes divergent-boundary earthquakes and produces numerous cliffs, or scarps, that lie parallel to the ridge axis.

How Does the Lithospheric Mantle Form at a Mid-Ocean Ridge?

So far, we've seen how oceanic *crust* forms at mid-ocean ridges. How does the *mantle* part of the oceanic lithosphere form? This part consists of the cooler uppermost layer of the mantle, in which temperatures are less than about 1,280°C. At the ridge axis, such temperatures occur almost at the base of the crust, because of the presence of rising hot asthenosphere

and hot magma, so lithospheric mantle beneath the ridge axis effectively doesn't exist. But as the newly formed oceanic crust moves away from the ridge axis, the crust and the uppermost mantle directly beneath it gradually cool by losing heat to the ocean above. As soon as mantle rock cools below 1,280°C, it becomes, by definition, part of the lithosphere.

As oceanic lithosphere continues to move away from the ridge axis, it continues to cool, so the lithospheric mantle, and therefore the oceanic lithosphere as a whole, grows progressively thicker (**Fig. 2.20a, b**). Note that this process doesn't change the thickness of the oceanic crust, for the crust formed entirely at the ridge axis. The rate at which cooling and lithospheric thickening occur decreases progressively with increasing distance from the ridge axis. In fact, by the time the lithosphere is about 80 million years old, it has just about reached its maximum thickness. As lithosphere thickens and gets cooler and denser, it sinks down into the asthenosphere, like a ship taking on ballast. Thus, the ocean is deeper over older ocean floor than over younger ocean floor.

FIGURE 2.17 The process of sea-floor spreading.

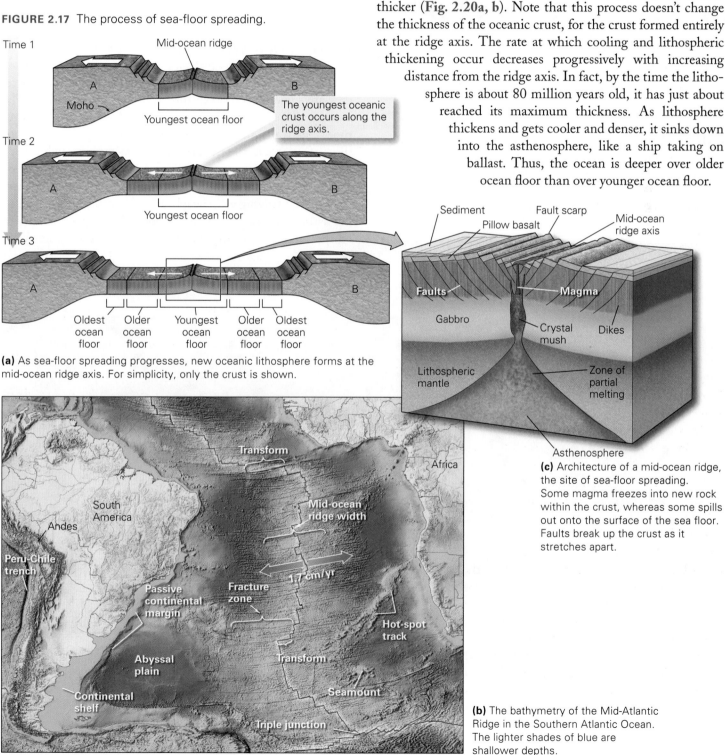

(a) As sea-floor spreading progresses, new oceanic lithosphere forms at the mid-ocean ridge axis. For simplicity, only the crust is shown.

(c) Architecture of a mid-ocean ridge, the site of sea-floor spreading. Some magma freezes into new rock within the crust, whereas some spills out onto the surface of the sea floor. Faults break up the crust as it stretches apart.

(b) The bathymetry of the Mid-Atlantic Ridge in the Southern Atlantic Ocean. The lighter shades of blue are shallower depths.

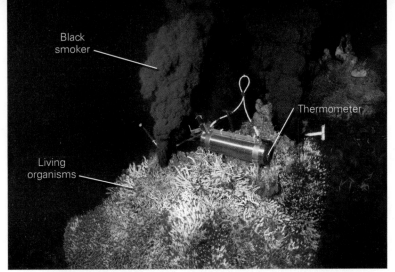

FIGURE 2.18 A column of superhot water gushing from a vent known as a black smoker along the mid-ocean ridge. A local ecosystem of bacteria, shrimp, and worms lives around the vent.

Take-Home **Message**

Sea-floor spreading occurs at divergent plate boundaries, defined by mid-ocean ridges. New oceanic crust solidifies from basaltic magma along the ridge axis. As plates move away from the axis, they cool, and the lithospheric mantle forms and thickens.

2.8 Convergent Plate Boundaries and Subduction

At convergent plate boundaries, two plates, at least one of which is oceanic, move toward one another. But rather than butting each other like angry rams, one oceanic plate bends and sinks down into the asthenosphere beneath the other plate. Geologists refer to the sinking process as **subduction**, so convergent boundaries are also known as subduction zones. Because subduction at a convergent boundary consumes old ocean lithosphere and thus "consumes" oceanic basins, geologists also refer to convergent boundaries as consuming boundaries, and because they are delineated by deep-ocean trenches, they are sometimes simply called **trenches** (see Fig. 2.8). The amount of oceanic plate consumption worldwide, averaged over time, equals the amount of sea-floor spreading worldwide, so the surface area of the Earth remains constant through time.

Subduction occurs for a simple reason: oceanic lithosphere, once it has aged at least 10 million years, is denser than the underlying asthenosphere and thus can sink through the asthenosphere if given an opportunity. Where it lies flat on the surface of the asthenosphere, oceanic lithosphere can't

FIGURE 2.19 This map of the world shows the age of the sea floor. Note how the sea floor grows older with increasing distance from the ridge axis. (Ma = million years ago.)

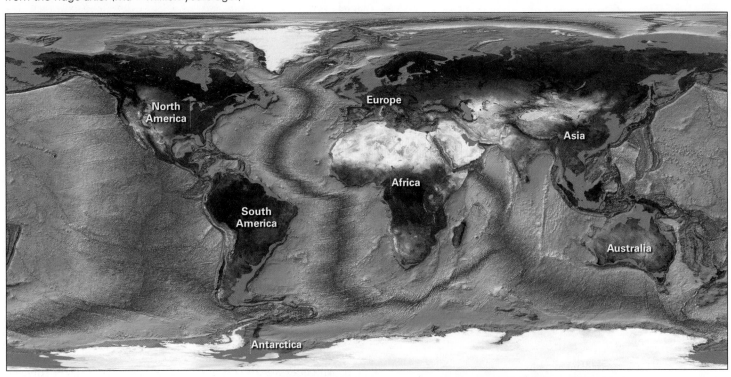

FIGURE 2.20 Changes accompanying the aging of lithosphere.

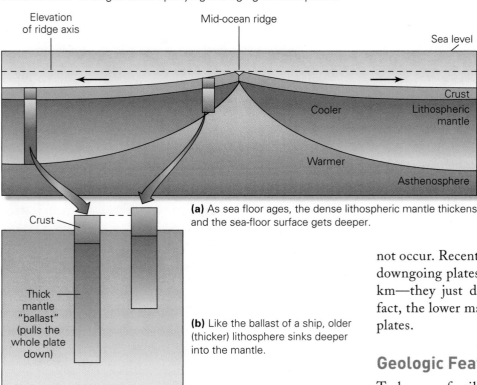

(a) As sea floor ages, the dense lithospheric mantle thickens and the sea-floor surface gets deeper.

(b) Like the ballast of a ship, older (thicker) lithosphere sinks deeper into the mantle.

sink. However, once the end of the convergent plate bends down and slips into the mantle, it continues downward like an anchor falling to the bottom of a lake (**Fig. 2.21a**). As the lithosphere sinks, asthenosphere flows out of its way, just as water flows out of the way of a sinking anchor. But unlike water, the asthenosphere can flow only very slowly, so oceanic lithosphere can sink only very slowly, at a rate of less than about 15 cm per year. To visualize the difference, imagine how much faster a coin can sink through water than it can through honey.

Note that the "downgoing plate," the plate that has been subducted, *must* be composed of oceanic lithosphere. The overriding plate, which does not sink, can consist of either oceanic or continental lithosphere. Continental crust cannot be subducted because it is too buoyant; the low-density rocks of continental crust act like a life preserver keeping the continent afloat. If continental crust moves into a convergent margin, subduction eventually stops. Because of subduction, all ocean floor on the planet is less than about 200 million years old. Because continental crust cannot subduct, some continental crust has persisted at the surface of the Earth for over 3.8 billion years.

Earthquakes and the Fate of Subducted Plates

At convergent plate boundaries, the downgoing plate grinds along the base of the overriding plate, a process that

generates large earthquakes. These earthquakes occur fairly close to the Earth's surface, so some of them cause massive destruction in coastal cities. But earthquakes also happen in downgoing plates at greater depths. In fact, geologists have detected earthquakes within downgoing plates to a depth of 660 km. The band of earthquakes in a downgoing plate is called a **Wadati-Benioff zone**, after its two discoverers (**Fig. 2.21b**).

At depths greater than 660 km, conditions leading to earthquakes in subducted lithosphere evidently do not occur. Recent observations, however, indicate that some downgoing plates do continue to sink *below* a depth of 660 km—they just do so without generating earthquakes. In fact, the lower mantle may be a graveyard for old subducted plates.

Geologic Features of a Convergent Boundary

To become familiar with the various geologic features that occur along a convergent plate boundary, let's look at an example, the boundary between the western coast of the South American Plate and the eastern edge of the Nazca Plate (a portion of the Pacific Ocean floor). A deep-ocean trench, the Peru-Chile Trench, delineates this boundary (see Fig. 2.17b). Such trenches form where the plate bends as it starts to sink into the asthenosphere.

In the Peru-Chile Trench, as the downgoing plate slides under the overriding plate, sediment (clay and plankton) that had settled on the surface of the downgoing plate, as well as sand that fell into the trench from the shores of South America, gets scraped up and incorporated in a wedge-shaped mass known as an **accretionary prism** (**Fig. 2.21c**). An accretionary prism forms in basically the same way as a pile of snow or sand in front of a plow, and like snow, the sediment tends to be squashed and contorted.

A chain of volcanoes known as a **volcanic arc** develops behind the accretionary prism (**See for Yourself B**). As we will see in Chapter 4, the magma that feeds these volcanoes forms just above the surface of the downgoing plate where the plate reaches a depth of about 150 km below the Earth's surface. If the volcanic arc forms where an oceanic plate subducts beneath continental lithosphere, the resulting chain of volcanoes grows on the continent and forms a continental volcanic arc. (In some cases, the plates squeeze together across a continental arc, causing a belt of faults to form behind the arc.) If, however, the volcanic arc grows

FIGURE 2.21 During the process of subduction, oceanic lithosphere sinks back into the deeper mantle.

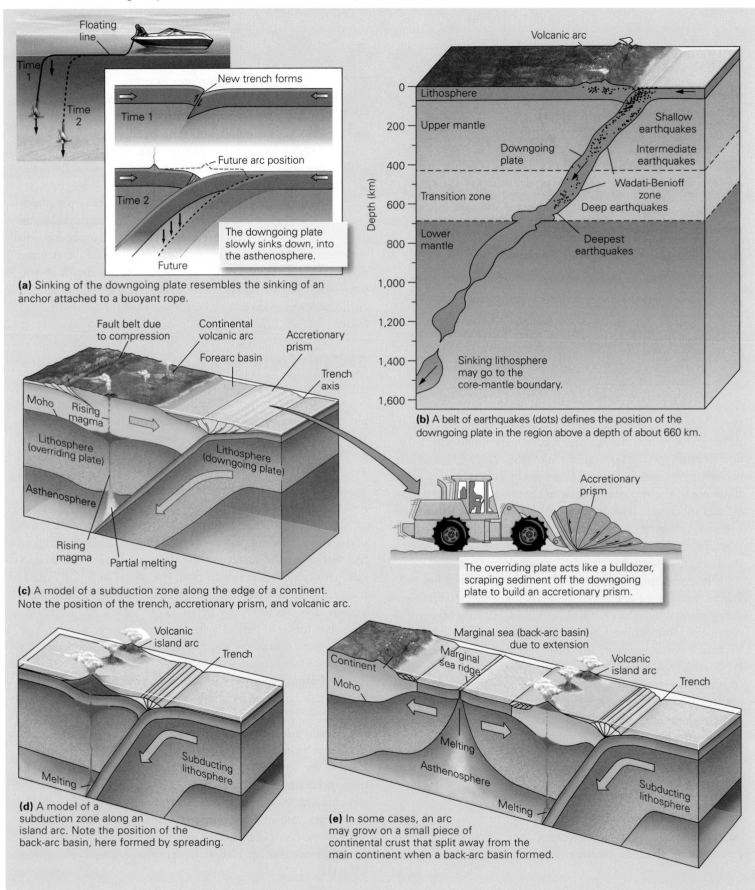

(a) Sinking of the downgoing plate resembles the sinking of an anchor attached to a buoyant rope.

(b) A belt of earthquakes (dots) defines the position of the downgoing plate in the region above a depth of about 660 km.

(c) A model of a subduction zone along the edge of a continent. Note the position of the trench, accretionary prism, and volcanic arc.

The overriding plate acts like a bulldozer, scraping sediment off the downgoing plate to build an accretionary prism.

(d) A model of a subduction zone along an island arc. Note the position of the back-arc basin, here formed by spreading.

(e) In some cases, an arc may grow on a small piece of continental crust that split away from the main continent when a back-arc basin formed.

where one oceanic plate subducts beneath another oceanic plate, the resulting volcanoes form a chain of islands known as a volcanic island arc (**Fig. 2.21d**). A back-arc basin exists either where subduction happens to begin offshore, trapping ocean lithosphere behind the arc, or where stretching of the lithosphere behind the arc leads to the formation of a small spreading ridge behind the arc (**Fig. 2.21e**).

> ## Take-Home **Message**
> At a convergent plate boundary, an oceanic plate sinks into the mantle beneath the edge of another plate. A volcanic arc and a trench delineate such plate boundaries, and earthquakes happen along the contact between the two plates as well as in the downgoing slab.

2.9 **Transform Plate Boundaries**

When researchers began to explore the bathymetry of mid-ocean ridges in detail, they discovered that mid-ocean ridges are not long, uninterrupted lines, but rather consist of short segments that appear to be offset laterally from each other (**Fig. 2.22a**) by narrow belts of broken and irregular sea floor. These belts, or **fracture zones**, lie roughly at right angles to the ridge segments, intersect the ends of the segments, and extend beyond the ends of the segments. Originally, researchers incorrectly assumed that the entire length of each fracture zone was a fault, and that slip on a fracture zone had displaced segments of the mid-ocean ridge sideways, relative to each other. In other words, they imagined that a mid-ocean ridge initiated as a continuous, fence-like line that only later was broken up by faulting. But when information about the distribution of earthquakes along mid-ocean ridges became available, it was clear that this model could not be correct. Earthquakes, and therefore active fault slip, occur *only* on the segment of a fracture zone that lies between two ridge segments. The portions of fracture zones that extend beyond the edges of ridge segments, out into the abyssal plain, are not seismically active.

The distribution of movement along fracture zones remained a mystery until a Canadian researcher, J. Tuzo Wilson, began to think about fracture zones in the context of the sea-floor-spreading concept. Wilson proposed that fracture zones formed *at the same time* as the ridge axis itself, and thus the ridge consisted of separate segments to start with. These segments were linked (not offset) by fracture zones. With this idea in mind, he drew a sketch map showing two ridge-axis segments linked by a fracture zone, and he drew arrows to indicate the direction that ocean floor was moving,

relative to the ridge axis, as a result of sea-floor-spreading (**Fig. 2.22b**). Look at the arrows in Figure 2.22b. Clearly, the movement direction on the active portion of the fracture zone must be opposite to the movement direction that researchers originally thought occurred on the structure. Further, in Wilson's model, slip occurs only along the segment of the fracture zone between the two ridge segments (**Fig. 2.22c**). Plates on opposite sides of the inactive part of a fracture zone move together, as one plate.

Wilson introduced the term **transform boundary**, or transform fault, for the actively slipping segment of a fracture zone between two ridge segments, and he pointed out that these are a third type of plate boundary. At a transform boundary, one plate slides sideways past another, but no new plate forms and no old plate is consumed. Transform boundaries are, therefore, defined by a vertical fault on which the slip direction parallels the Earth's surface. The slip breaks up the crust and forms a set of steep fractures.

So far we've discussed only transforms along mid-ocean ridges. Not all transforms link ridge segments. Some, such as the Alpine Fault of New Zealand, link trenches, while others link a trench to a ridge segment. Further, not all transform faults occur in oceanic lithosphere; a few cut across continental lithosphere. The San Andreas Fault, for example, which cuts across California, defines part of the plate boundary between the North American Plate and the Pacific Plate—the portion of California that lies to the west of the fault (including Los Angeles) is part of the Pacific Plate, while the portion that lies to the east of the fault is part of the North American Plate (**Fig. 2.22d, e**).

> ## Take-Home **Message**
> At transform plate boundaries, one plate slips sideways past another. Most transform boundaries link segments of mid-ocean ridges, but some, such as the San Andreas Fault, cut across continental crust.

2.10 **Special Locations in the Plate Mosaic**

Triple Junctions

Geologists refer to a place where three plate boundaries intersect as a **triple junction**, and name them after the types of boundaries that intersect. For example, the triple junction formed where the Southwest Indian Ocean Ridge intersects two arms of the Mid–Indian Ocean Ridge (this is the triple

FIGURE 2.22 The concept of transform faulting.

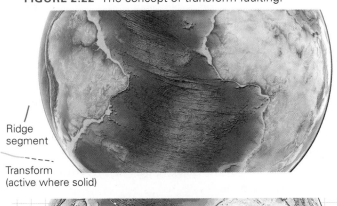

Ridge
segment

Transform
(active where solid)

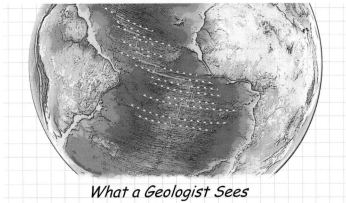

What a Geologist Sees

(a) Transform faults segment the Mid-Atlantic Ridge.

Old idea (incorrect)   **New idea (correct)**

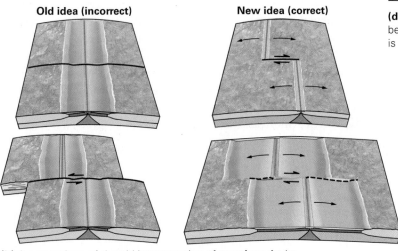

(b) A comparison of the old interpretation of transform faults with the new interpretation required by the sea-floor spreading hypothesis. Note that the motion on the transform must be compatible with the direction of spreading.

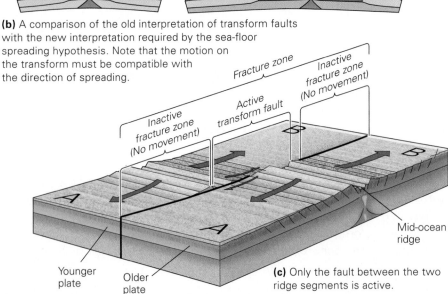

Fracture zone

Inactive
fracture zone
(No movement)

Active
transform fault

Inactive
fracture zone
(No movement)

Younger
plate

Older
plate

Mid-ocean
ridge

(c) Only the fault between the two ridge segments is active.

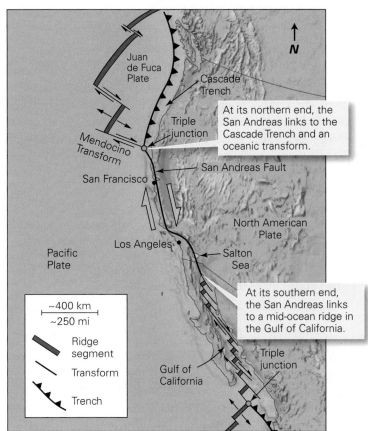

Juan
de Fuca
Plate

Cascade
Trench

Triple
junction

Mendocino
Transform

At its northern end, the
San Andreas links to the
Cascade Trench and an
oceanic transform.

San Francisco

San Andreas Fault

Los Angeles

North American
Plate

Pacific
Plate

Salton
Sea

At its southern end,
the San Andreas links
to a mid-ocean ridge in
the Gulf of California.

~400 km
~250 mi

Ridge
segment

Transform

Trench

Gulf of
California

Triple
junction

(d) The San Andreas Fault is a transform plate boundary between the North American and Pacific Plates. The Pacific is moving northwest, relative to North America.

Fault
trace

(e) In southern California, the San Andreas Fault cuts a dry landscape. The fault trace is in the narrow valley. The land has been pushed up slightly along the fault.

FIGURE 2.23 Examples of triple junctions. The triple junctions are marked by dots.

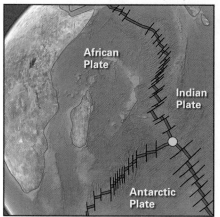

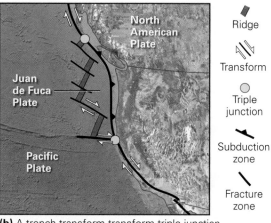

(a) A ridge-ridge-ridge triple junction occurs in the Indian Ocean.

(b) A trench-transform-transform triple junction occurs at the north end of the San Andreas Fault.

junction of the African, Antarctic, and Australian Plates) is a ridge-ridge-ridge triple junction (**Fig. 2.23a**). The triple junction north of San Francisco is a trench-transform-transform triple junction (**Fig. 2.23b**).

Hot Spots

Most subaerial (above sea level) volcanoes are situated in the volcanic arcs that border trenches. Volcanoes also lie along mid-ocean ridges, but ocean water hides most of them. The volcanoes of volcanic arcs and mid-ocean ridges *are plate-boundary volcanoes*, in that they formed as a consequence of movement along the boundary. Not all volcanoes on Earth are plate-boundary volcanoes, however. Worldwide, geoscientists have identified about 100 volcanoes that exist as isolated points and are not a consequence of movement at a plate boundary. These are called hot-spot volcanoes, or simply **hot spots** (**Fig. 2.24**). Most hot spots are located in the interiors of plates, away from the boundaries, but a few lie along mid-ocean ridges.

What causes hot-spot volcanoes? In the early 1960s, J. Tuzo Wilson noted that *active* hot-spot volcanoes (examples that are erupting or may erupt in the future) occur at the end of a chain of *dead* volcanic islands and seamounts (formerly active volcanoes that will never erupt again). This configuration is different from that of volcanic arcs along convergent plate boundaries—at volcanic arcs, all of the volcanoes are active. With this image in mind, Wilson suggested that the position of the heat source causing a hot-spot volcano is fixed, relative to the moving plate. In Wilson's model, the active volcano represents the present-day location of the heat source, whereas the chain of dead volcanic islands represents locations on the plate that were once over the heat source but progressively moved off.

A few years later, researchers suggested that the heat source for hot spots is a **mantle plume**, a column of very hot rock rising up through the mantle to the base of the lithosphere (**Fig. 2.25a–d**). In this model, plumes originate deep in the mantle. Rock in the plume, though solid, is soft

Legend:
Ridge
Transform
Triple junction
Subduction zone
Fracture zone

FIGURE 2.24 The dots represent the locations of selected hot-spot volcanoes. The red lines represent hot-spot tracks. The most recent volcano (dot) is at one end of this track. Some of these volcanoes are extinct, indicating that the mantle plume no longer exists. Some hot spots are fairly recent and do not have tracks. Dashed tracks were broken by sea-floor spreading.

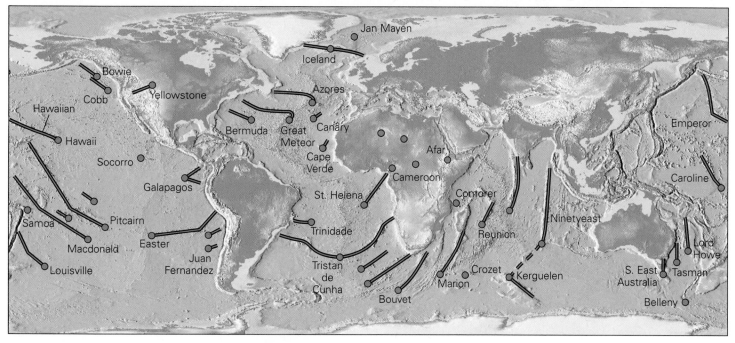

FIGURE 2.25 The deep mantle plume hypothesis for the formation of hot-spot tracks.

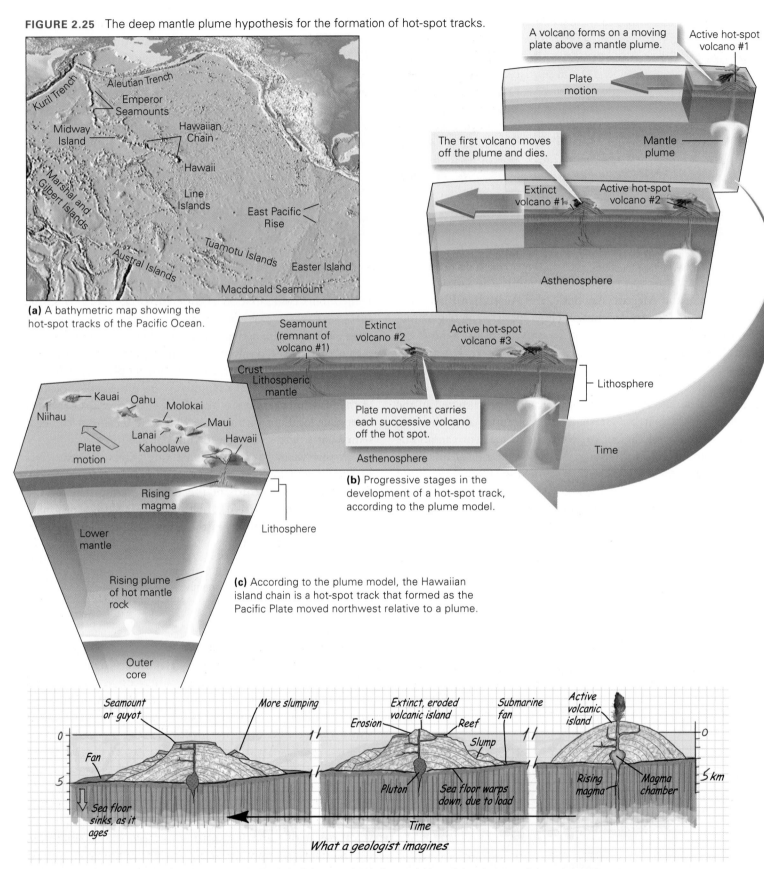

(a) A bathymetric map showing the hot-spot tracks of the Pacific Ocean.

A volcano forms on a moving plate above a mantle plume.

Active hot-spot volcano #1

Plate motion

Mantle plume

The first volcano moves off the plume and dies.

Extinct volcano #1

Active hot-spot volcano #2

Asthenosphere

Seamount (remnant of volcano #1)

Extinct volcano #2

Active hot-spot volcano #3

Crust
Lithospheric mantle

Lithosphere

Plate movement carries each successive volcano off the hot spot.

Asthenosphere

Time

(b) Progressive stages in the development of a hot-spot track, according to the plume model.

Kauai Oahu Molokai
Niihau   Maui
Plate   Lanai   Hawaii
motion   Kahoolawe

Rising magma

Lithosphere

Lower mantle

Rising plume of hot mantle rock

(c) According to the plume model, the Hawaiian island chain is a hot-spot track that formed as the Pacific Plate moved northwest relative to a plume.

Outer core

Seamount or guyot   More slumping   Extinct, eroded volcanic island   Submarine fan   Active volcanic island

Erosion   Reef

Fan   Slump

Pluton   Sea floor warps down, due to load   Rising magma   Magma chamber

Sea floor sinks, as it ages   Time

What a geologist imagines

(d) As a volcano moves off the hot spot, it gradually sinks below sea level, due to sinking of the plate, erosion, and slumping.

enough to flow, and rises buoyantly because it is less dense than surrounding cooler rock. When the hot rock of the plume reaches the base of the lithosphere, it partially melts (for reasons discussed in Chapter 4) and produces magma that seeps up through the lithosphere to the Earth's surface. The chain of extinct volcanoes, or **hot-spot track**, forms when the overlying plate moves over a fixed plume. This movement slowly carries the volcano off the top of the plume, so that it becomes extinct. A new, younger volcano grows over the plume.

The Hawaiian chain provides an example of the volcanism associated with a hot-spot track. Volcanic eruptions occur today only on the big island of Hawaii. Other islands to the northwest are remnants of dead volcanoes, the oldest of which is Kauai. To the northwest of Kauai, still older volcanic remnants are found. About 1,750 km northwest of Midway Island, the track bends in a more northerly direction, and the volcanic remnants no longer poke above sea level; we refer to this northerly trending segment as the Emperor seamount chain. Geologists suggest that the bend is due to a change in the direction of Pacific Plate motion at about 40 Ma.

Some hot spots lie within continents. For example, several have been active in the interior of Africa, and one now underlies Yellowstone National Park. The famous geysers (natural steam and hot-water fountains) of Yellowstone exist because hot magma, formed above the Yellowstone hot spot, lies not far below the surface of the park. While most hot spots, such as Hawaii and Yellowstone, occur in the interior of plates, away from plate boundaries, a few are positioned at points on mid-ocean ridges. The additional magma production associated with such hot spots causes a portion of the ridge to grow into a mound that can rise significantly above normal ridge-axis depths and protrude above the sea surface. Iceland, for example, is the product of hot-spot volcanism on the axis of the Mid-Atlantic Ridge.

> ### Take-Home Message
>
> A triple junction marks the point where three plate boundaries join. A hot spot is a place where volcanism may be due to melting at the top of a mantle plume. As a plate moves over a plume, a hot-spot track develops.

2.11 How Do Plate Boundaries Form and Die?

The configuration of plates and plate boundaries visible on our planet today has not existed for all of geologic history, and will not exist indefinitely into the future. Because of plate motion, oceanic plates form and are later consumed, while continents merge and later split apart. How does a new divergent boundary come into existence, and how does an existing convergent boundary eventually cease to exist? Most new divergent boundaries form when a continent splits and separates into two continents. We call this process **rifting**. A convergent boundary ceases to exist when a piece of buoyant lithosphere, such as a continent or an island arc, moves into the subduction zone and, in effect, jams up the system. We call this process **collision**.

Continental Rifting

A **continental rift** is a linear belt in which continental lithosphere pulls apart (**Fig. 2.26a**). During the process, the lithosphere stretches horizontally and thins vertically, much like a piece of taffy you pull between your fingers. Nearer the surface of the continent, where the crust is cold and brittle, stretching causes rock to break and faults to develop. Blocks of rock slip down the fault surfaces, leading to the formation of a low area that gradually becomes buried by sediment. Deeper in the crust, and in the underlying lithospheric mantle, rock is warmer and softer, so stretching takes place in a plastic manner without breaking the rock. The whole region that stretches is the rift, and the process of stretching is called rifting.

As continental lithosphere thins, hot asthenosphere rises beneath the rift and starts to melt. Eruption of the molten rock produces volcanoes along the rift. If rifting continues for a long enough time, the continent breaks in two, a new mid-ocean ridge forms, and sea-floor spreading begins. The relict of the rift evolves into a passive margin (see Fig. 2.14b). In some cases, however, rifting stops before the continent splits in two; it becomes a low-lying trough that fills with sediment. Then, the rift remains as a permanent scar in the crust, defined by a belt of faults, volcanic rocks, and a thick layer of sediment.

A major rift, known as the Basin and Range Province, breaks up the landscape of the western United States (**Fig. 2.26b**). Here, movement on numerous faults tilted blocks of crust to form narrow mountain ranges, while sediment that eroded from the blocks filled the adjacent basins (the low areas between the ranges). Another active rift slices through eastern Africa; geoscientists aptly refer to it as the East African Rift (**Fig. 2.26c, d**). To astronauts in orbit, the rift looks like a giant gash in the crust. On the ground, it consists of a deep trough bordered on both sides by high cliffs formed by faulting. Along the length of the rift, several major volcanoes smoke and fume; these include the snow-crested Mt. Kilimanjaro, towering over 6 km above the savannah. At its north end, the rift joins the Red Sea Ridge and the Gulf of Aden Ridge at a triple junction.

FIGURE 2.26 During the process of rifting, lithosphere stretches.

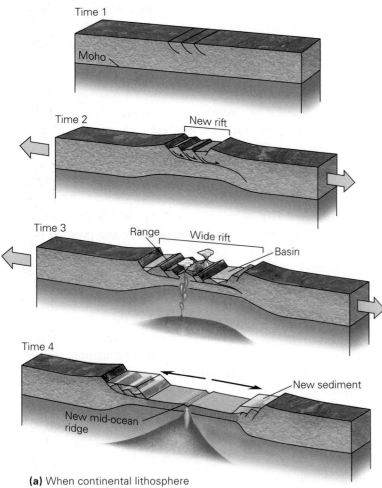

(a) When continental lithosphere stretches and thins, faulting takes place, and volcanoes erupt. Eventually, the continent splits in two and a new ocean basin forms.

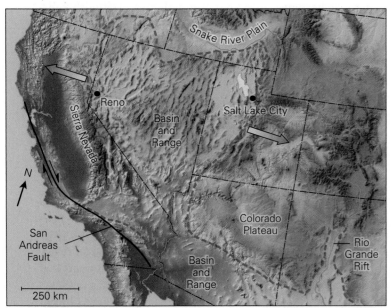

(b) The Basin and Range Province is a rift. Faulting bounds the narrow north-south-trending mountains, separated by basins. The arrows indicate the direction of stretching.

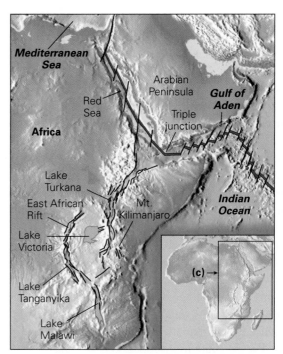

(c) The East African Rift is growing today. The Red Sea started as a rift. The inset shows map locations.

(d) An air photo of the northern end of the East African rift, showing faults and volcanoes.

Collision

India was once a small, separate continent that lay far to the south of Asia. But subduction consumed the ocean between India and Asia, and India moved northward, finally slamming into the southern margin of Asia about 40 to 50 million years ago. Continental crust, unlike oceanic crust, is too buoyant to subduct. So when India collided with Asia, the attached oceanic plate broke off and sank down into the deep mantle while India pushed hard into and partly under Asia, squeezing the rocks and sediment that once lay between the two continents into the 8-km-high welt that we now know as the Himalayan Mountains. During this process, not only did the surface of the Earth rise, but the crust became thicker. The crust beneath a collisional mountain range can be up to 60 to 70 km thick, about twice the thickness of normal continental crust. The boundary between what was once two separate continents is called a suture; slivers of ocean crust may be trapped along a suture.

Geoscientists refer to the process during which two buoyant pieces of lithosphere converge and squeeze together as collision (**Fig. 2.27a, b**). Some collisions involve two continents, whereas some involve continents and an island arc. When a collision is complete, the convergent plate boundary that once existed between the two colliding pieces ceases to exist. Collisions yield some of the most spectacular mountains on the planet, such as the Himalayas and the Alps. They also yielded major mountain ranges in the past, which subsequently eroded away so that today we see only

FIGURE 2.27 Continental collision (not to scale).

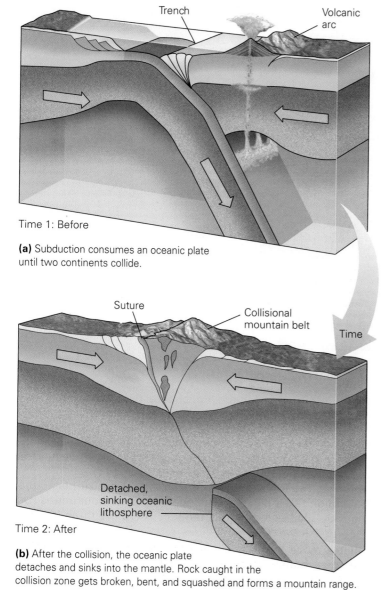

Time 1: Before

(a) Subduction consumes an oceanic plate until two continents collide.

Time 2: After

(b) After the collision, the oceanic plate detaches and sinks into the mantle. Rock caught in the collision zone gets broken, bent, and squashed and forms a mountain range.

their relicts. For example, the Appalachian Mountains in the eastern United States formed as a consequence of three collisions. After the last one, a collision between Africa and North America around 300 Ma, North America became part of the Pangaea supercontinent.

Take-Home Message

Rifting can split a continent in two and can lead to the formation of a new divergent plate boundary. When two buoyant crustal blocks, such as continents and island arcs, collide, a mountain belt forms and subduction eventually ceases.

2.12 What Drives Plate Motion, and How Fast Do Plates Move?

Forces Acting on Plates

We've now discussed the many facets of plate tectonics theory (see **Geology at a Glance**, pp. 64–65). But to complete the story, we need to address a major question: "What drives plate motion?" When geoscientists first proposed plate tectonics, they thought the process occurred simply because convective flow in the asthenosphere actively dragged plates along, as if the plates were simply rafts on a flowing river. Thus, early images depicting plate motion showed simple convection cells—elliptical flow paths—in the asthenosphere. At first glance, this hypothesis looked pretty good. But, on closer examination it became clear that a model of simple convection cells carrying plates on their backs can't explain the complex geometry of plate boundaries and the great variety of plate motions that we observe on the Earth. Researchers now prefer a model in which convection, ridge push, and slab pull all contribute to driving plates. Let's look at each of these phenomena in turn.

Convection is involved in plate motions in two ways. Recall that, at a mid-ocean ridge, hot asthenosphere rises and then cools to form oceanic lithosphere which slowly moves away from the ridge until, eventually, it sinks back into the mantle at a trench. Since the material forming the plate starts out hot, cools, and then sinks, we can view the plate itself as the top of a convection cell and plate motion as a form of convection. But in this view, convection is effectively a consequence of plate motion, not the cause. Can convection actually cause plates to move? The answer may come from studies which demonstrate that the interior of the mantle, beneath the plates, is indeed convecting on a very broad scale (see Interlude D). Specifically, geologists have found that there are places where deeper, hotter asthenosphere is rising or *upwelling*, and places where shallower, colder asthenosphere is sinking or *downwelling*. Such

Theory of Plate Tectonics

Hot-spot
volcano

Transform-
plate
boundary

Volcanic arc

Trench

Continental rift

Convergent plate
boundary

Subducting oceanic
lithosphere

Collisional mountain belt

Continental crust

Continental lithosphere

Lithospheric mantle

Asthenosphere

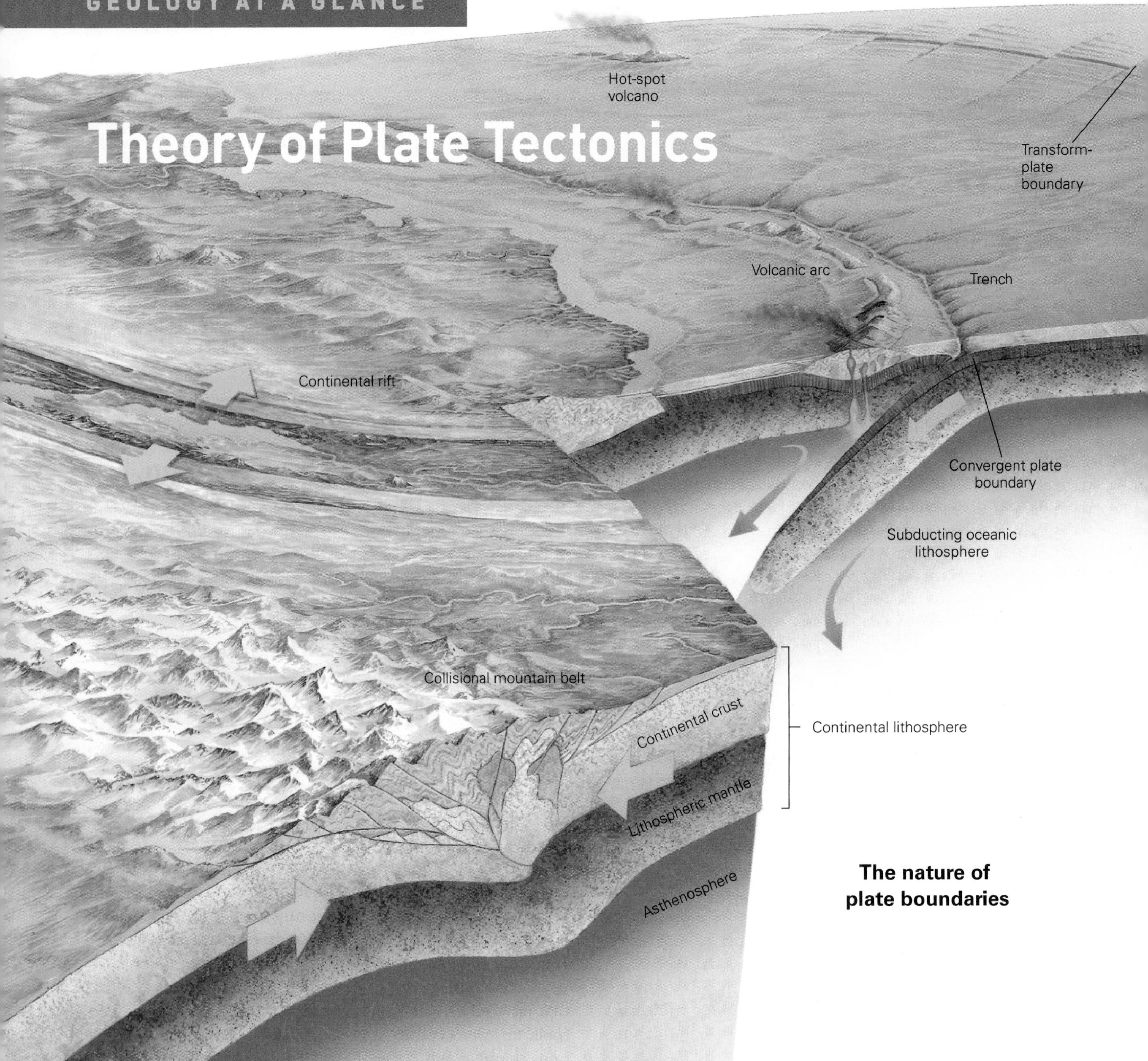

**The nature of
plate boundaries**

The outer portion of the Earth is a relatively rigid layer called the lithosphere. It consists of the crust (oceanic or continental) and the uppermost mantle. The mantle below the lithosphere is relatively plastic (it can flow) and is called the asthenosphere. The difference in behavior (rigid vs. plastic) between lithospheric mantle and asthenospheric mantle is a consequence of temperature—the former is cooler than the latter. Continental lithosphere is typically about 150 km thick, while oceanic lithosphere is about 100 km thick. (*Note*: These are not drawn to scale in this image.)

According to the theory of plate tectonics, the lithosphere is broken into about 20 plates that move relative to each other. Most of the motion takes place by sliding along plate boundaries (the edges of plates); plate interiors stay relatively unaffected by this motion. There are three kinds of plate boundaries.

1. Divergent boundaries: Here, two plates move apart by a process called sea-floor spreading. A mid-ocean ridge delineates a divergent boundary. Asthenospheric mantle rises beneath a

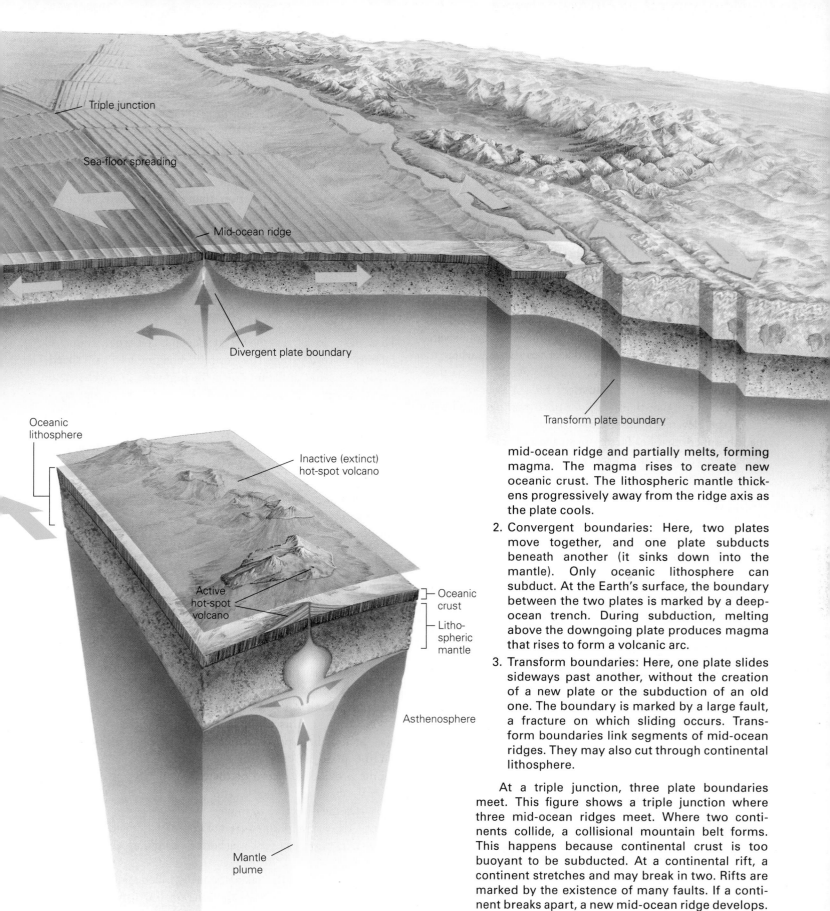

Triple junction

Sea-floor spreading

Mid-ocean ridge

Divergent plate boundary

Oceanic lithosphere

Inactive (extinct) hot-spot volcano

Active hot-spot volcano

Oceanic crust

Lithospheric mantle

Asthenosphere

Mantle plume

Formation of a hot-spot track

Transform plate boundary

mid-ocean ridge and partially melts, forming magma. The magma rises to create new oceanic crust. The lithospheric mantle thickens progressively away from the ridge axis as the plate cools.

2. Convergent boundaries: Here, two plates move together, and one plate subducts beneath another (it sinks down into the mantle). Only oceanic lithosphere can subduct. At the Earth's surface, the boundary between the two plates is marked by a deep-ocean trench. During subduction, melting above the downgoing plate produces magma that rises to form a volcanic arc.

3. Transform boundaries: Here, one plate slides sideways past another, without the creation of a new plate or the subduction of an old one. The boundary is marked by a large fault, a fracture on which sliding occurs. Transform boundaries link segments of mid-ocean ridges. They may also cut through continental lithosphere.

At a triple junction, three plate boundaries meet. This figure shows a triple junction where three mid-ocean ridges meet. Where two continents collide, a collisional mountain belt forms. This happens because continental crust is too buoyant to be subducted. At a continental rift, a continent stretches and may break in two. Rifts are marked by the existence of many faults. If a continent breaks apart, a new mid-ocean ridge develops.

Hot-spot volcanoes may form above plumes of hot mantle rock that rise from near the core-mantle boundary. As a plate drifts over a hot spot, it leaves a chain of extinct volcanoes.

asthenospheric flow probably does exert a force on the base of plates. But the pattern of upwelling and downwelling on a global scale does not match the pattern of plate boundaries exactly. So, conceivably, asthenosphere-flow may either speed up or slow down plates depending on the orientation of the flow direction relative to the movement direction of the overlying plate.

Ridge-push force develops simply because the lithosphere of mid-ocean ridges lies at a higher elevation than that of the adjacent abyssal plains (**Fig. 2.28a**). To understand ridge-push force, imagine you have a glass containing a layer of water over a layer of honey. By tilting the glass momentarily and then returning it to its upright position, you can create a temporary slope in the boundary between these substances. While the boundary has this slope, gravity causes the weight of elevated honey to push against the glass adjacent to the side where the honey surface lies at lower elevation. The geometry of a mid-ocean ridge resembles this situation, for sea floor of a mid-ocean ridge is higher than sea floor of abyssal plains. Gravity causes the elevated lithosphere at the ridge axis to push on the lithosphere that lies farther from the axis, making it move away. As lithosphere moves away from the ridge axis, new hot

asthenosphere rises to fill the gap. Note that the local upward movement of asthenosphere beneath a mid-ocean ridge is a *consequence* of sea-floor spreading, not the cause.

Slab-pull force, the force that subducting, downgoing plates apply to oceanic lithosphere at a convergent margin, arises simply because lithosphere that was formed more than 10 million years ago is denser than asthenosphere, so it can sink into the asthenosphere (**Fig. 2.28b**). Thus, once an oceanic plate starts to sink, it gradually pulls the rest of the plate along behind it, like an anchor pulling down the anchor line. This "pull" is the slab-pull force.

The Velocity of Plate Motions

How fast do plates move? It depends on your frame of reference. To illustrate this concept, imagine two cars speeding in the same direction down the highway. From the viewpoint of a tree along the side of the road, Car A zips by at 100 km an hour, while Car B moves at 80 km an hour. But relative to Car B, Car A moves at only 20 km an hour. Geologists use two different frames of reference for describing plate velocity. If we describe the movement of Plate A with respect to Plate B, then we are speaking about **relative plate velocity**. But if we describe the movement of both plates relative to a fixed location in the mantle below the plates, then we are speaking of **absolute plate velocity** (**Fig. 2.29**).

To determine relative plate motions, geoscientists measure the distance of a known magnetic anomaly from the axis of a mid-ocean ridge and then calculate the velocity of a plate relative to the ridge axis by applying this equation: plate velocity = distance from the anomaly to the ridge axis divided by the age of the anomaly (velocity, by definition, is distance ÷ time). The velocity of the plate on one side of the ridge relative to the plate on the other is twice this value.

To estimate absolute plate motions, we can *assume* that the position of a mantle plume does not change much for a long time. If this is so, then the track of hot-spot volcanoes on the plate moving over the plume provides a record of the plate's absolute velocity and indicates the direction of movement. (In reality, plumes are not completely fixed; geologists use other, more complex methods to calculate absolute plate motions.)

> **Did you ever wonder...**
> whether we can really "see" continents drift?

Working from the calculations described above, geologists have determined that plate motions on Earth today occur at rates of about 1 to 15 cm per year—about the rate that your fingernails grow. But these rates, though small, can yield large displacements given the immensity of geologic time. At a rate of 10 cm per year, a plate can move 100 km in a million years! Can we detect such slow rates? Until the last decade, the

FIGURE 2.28 Forces driving plate motions. Both ridge push and slab pull make plates move.

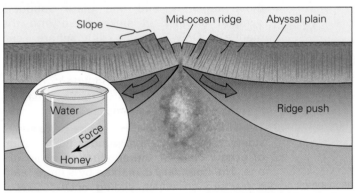

(a) Ridge push develops because the region of a rift is elevated. Like a wedge of honey with a sloping surface, the mass of the ridge pushes sideways.

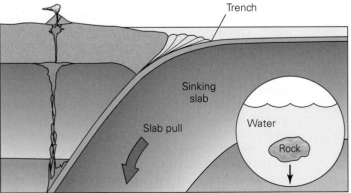

(b) Slab pull develops because lithosphere is denser than the underlying asthenosphere, and sinks like a stone in water (though much more slowly).

FIGURE 2.29 *Relative plate velocities*: The blue arrows show the rate and direction at which the plate on one side of the boundary is moving with respect to the plate on the other side. The length of an arrow represents the velocity. *Absolute plate velocities*: The red arrows show the velocity of the plates with respect to a fixed point in the mantle.

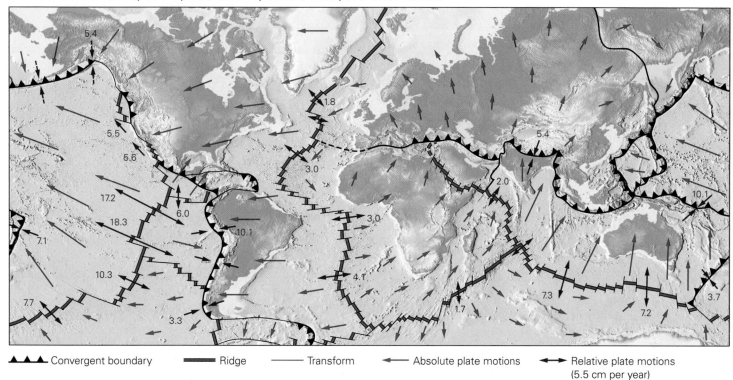

▲▲▲ Convergent boundary   ▬▬ Ridge   ▬▬ Transform   ◄— Absolute plate motions   ◄—► Relative plate motions (5.5 cm per year)

answer was no. Now the answer is yes, because of satellites orbiting the Earth with **global positioning system (GPS)** technology. Automobile drivers use GPS receivers to find their destinations, and geologists use them to monitor plate motions. If we calculate carefully enough, we can detect displacements of millimeters per year. In other words, we can now see the plates move—this observation serves as the ultimate proof of plate tectonics.

Taking into account many data sources that define the motion of plates, geologists have greatly refined the image of continental drift that Wegener tried so hard to prove nearly a century ago. We can now see how the map of our planet's surface has evolved radically during the past 400 million years (**Fig. 2.30**), and even before.

Take-Home Message

Plates move at 1 to 15 cm/yr. Relative motion specifies the rate that a plate moves relative to its neighbor, whereas absolute motion specifies the rate that a plate moves relative to a fixed point beneath the plate. GPS measurements can detect relative plate motions directly.

FIGURE 2.30 Due to plate tectonics, the map of Earth's surface slowly changes. Here we see the assembly, and later the breakup, of Pangaea during the past 400 million years.

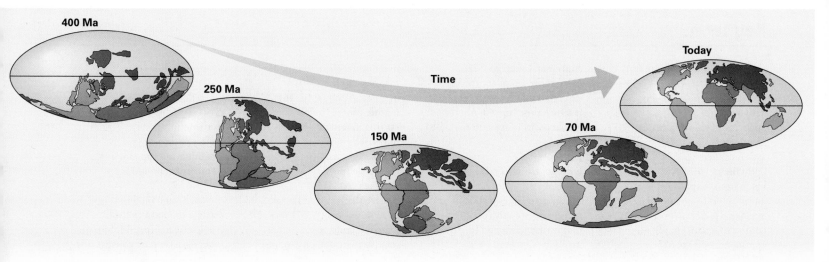

CHAPTER 2 REVIEW

Chapter Summary

> Alfred Wegener proposed that continents had once been joined together to form a single huge supercontinent (Pangaea) and had subsequently drifted apart. This idea is the continental drift hypothesis.

> Wegener drew from several different sources of data to support his hypothesis: (1) the correlation of coastlines; (2) the distribution of late Paleozoic glaciers; (3) the distribution of late Paleozoic climatic belts; (4) the distribution of fossil species; and (5) correlation of distinctive rock assemblages now on opposite sides of the ocean.

> Rocks retain a record of the Earth's magnetic field that existed at the time the rocks formed. This record is called paleomagnetism. By measuring paleomagnetism in successively older rocks, geologists discovered apparent polar-wander paths.

> Apparent polar-wander paths are different for different continents, because continents move with respect to each other, while the Earth's magnetic poles remain roughly fixed.

> Around 1960, Harry Hess proposed the hypothesis of sea-floor spreading. According to this hypothesis, new sea floor forms at mid-ocean ridges, then spreads symmetrically away from the ridge axis. Eventually, the ocean floor sinks back into the mantle at deep-ocean trenches.

> Geologists documented that the Earth's magnetic field reverses polarity every now and then. The record of reversals is called the magnetic-reversal chronology.

> A proof of sea-floor spreading came from the interpretation of marine magnetic anomalies and from drilling studies which proved that sea floor gets progressively older away from a mid-ocean ridge.

> The lithosphere is broken into discrete plates that move relative to each other. Continental drift and sea-floor spreading are manifestations of plate movement.

> Most earthquakes and volcanoes occur along plate boundaries; the interiors of plates remain relatively rigid and intact.

> There are three types of plate boundaries—divergent, convergent, and transform—distinguished from each other by the movement the plate on one side of the boundary makes relative to the plate on the other side.

> Divergent boundaries are marked by mid-ocean ridges. At divergent boundaries, sea-floor spreading produces new oceanic lithosphere.

> Convergent boundaries are marked by deep-ocean trenches and volcanic arcs. At convergent boundaries, oceanic lithosphere subducts beneath an overriding plate.

> Transform boundaries are marked by large faults at which one plate slides sideways past another. No new plate forms and no old plate is consumed at a transform boundary.

> Triple junctions are points where three plate boundaries intersect.

> Hot spots are places where volcanism occurs at an isolated volcano. As a plate moves over the hot spot, the volcano moves off and dies, and a new volcano forms over the hot spot. Hot spots may be caused by mantle plumes.

> A large continent can split into two smaller ones by the process of rifting. During rifting, continental lithosphere stretches and thins. If it finally breaks apart, a new mid-ocean ridge forms and sea-floor spreading begins.

> Convergent plate boundaries cease to exist when a buoyant piece of crust (a continent or an island arc) moves into the subduction zone. When that happens, collision occurs.

> Ridge-push force and slab-pull force contribute to driving plate motions. Plates move at rates of about 1 to 15 cm per year. Modern satellite measurements can detect these motions.

Key Terms

absolute plate velocity (p. 66)
abyssal plain (p. 43)
accretionary prism (p. 55)
active margin (p. 51)
apparent polar-wander path (p. 42)
asthenosphere (p. 50)
bathymetry (p. 43)
black smoker (p. 52)
chron (p. 47)
collision (p. 61)
continental drift (p. 35)
continental rift (p. 61)

continental shelf (p. 52)
convergent boundary (p. 52)
divergent boundary (p. 52)
fracture zone (pp. 43, 57)
global positioning system (GPS) (p. 67)
hot spot (p. 59)
hot-spot track (p. 61)
lithosphere (p. 50)
lithosphere plate (p. 51)
magnetic anomaly (p. 46)
magnetic declination (p. 40)
magnetic dipole (p. 40)

magnetic inclination (p. 41)
magnetic pole (p. 40)
magnetic reversal (p. 46)
mantle plume (p. 59)
marine magnetic anomaly (p. 46)
mid-ocean ridge (p. 43)
paleomagnetism (p. 41)
paleopole (p. 42)
Pangaea (p. 35)
passive margin (p. 51)
plate boundary (p. 51)
plate tectonics (p. 36)
relative plate velocity (p. 66)

ridge-push force (p. 66)
rifting (p. 61)
sea-floor spreading (pp. 36, 45)
seamount (p. 43)
slab-pull force (p. 66)
subduction (pp. 36, 54)
transform boundary (pp. 52, 57)
trench (pp. 43, 54)
triple junction (p. 57)
volcanic arc (pp. 43, 55)
Wadati-Benioff zone (p. 55)

Every chapter of SmartWork contains active learning exercises to assist you with reading comprehension and concept mastery. This chapter also features:

> Animation exercises on plate movements, subduction, and hot spots.

> A video exercise on divergent plate boundaries.
> Problems that help students determine relative plate velocities.

Review Questions

1. What was Wegener's continental drift hypothesis? What was his evidence? Why didn't other geologists agree?

2. How do apparent polar-wander paths show that the continents, rather than the poles, have moved?

3. Describe the hypothesis of sea-floor spreading.

4. Describe the pattern of marine magnetic anomalies across a mid-ocean ridge. How is this pattern explained?

5. How did drilling into the sea floor contribute further proof of sea-floor spreading? How did the sea-floor-spreading hypothesis explain variations in ocean floor heat flow?

6. What are the characteristics of a lithosphere plate? Can a single plate include both continental and oceanic lithosphere?

7. How does oceanic lithosphere differ from continental lithosphere in thickness, composition, and density?

8. How do we identify a plate boundary?

9. Describe the three types of plate boundaries.

10. How does crust form along a mid-ocean ridge?

11. Why is the oldest oceanic lithosphere less than 200 Ma?

12. Describe the major features of a convergent boundary.

13. Why are transform plate boundaries required on an Earth with spreading and subducting plate boundaries?

14. What is a triple junction?

15. How is a hot-spot track produced, and how can hot-spot tracks be used to track the past motions of a plate?

16. Describe the characteristics of a continental rift and give examples of where this process is occurring today.

17. Describe the process of continental collision and give examples of where this process has occurred.

18. Discuss the major forces that move lithosphere plates.

19. Explain the difference between relative plate velocity and absolute plate velocity.

On Further Thought

20. Why are the marine magnetic anomalies bordering the East Pacific Rise in the Pacific Ocean wider than those bordering the Mid-Atlantic Ridge?

21. The North Atlantic Ocean is 3,600 km wide. Sea-floor spreading along the Mid-Atlantic Ridge occurs at 2 cm per year. When did rifting start to open the Atlantic?

SEE FOR YOURSELF B...   Plate Tectonics

Download the *Google Earth*™ from the Web in order to visit the locations described below (instructions appear in the Preface of this book). You'll find further locations and associated active-learning exercises on Worksheet B of our **Geotours Workbook**.

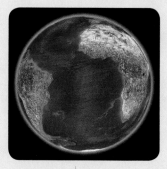

The South Atlantic
Latitude   11°38′14.93″S,
Longitude   14°12′57.84″W
A view of the South Atlantic from 13,500 km emphasizes that South America's coast looks like it could fit tightly against Africa's. If you rotate the Earth, you'll see that the east coast of the United States could fit against Africa's north-west coast.

Triple Junction, Japan
Latitude   37°56′27.58″N,
Longitude   140°28′53.87″E
A space view of Japan's coast from 3,250 km shows the presence of deep-sea trenches and a broad accretionary prism. The Pacific is subducting beneath Japan. A triple junction of three trenches lies along the east coast.

CHAPTER

3

Patterns in Nature: Minerals

Chapter Objectives

By the end of this chapter, you should know . . .

> the special meaning of the word "mineral" when used in a geologic context.

> how to organize the thousands of different minerals into just a few classes based on the chemicals the minerals contain.

> which minerals are the most common ones on Earth, and thus serve as the main building blocks of this planet.

> how to identify common mineral specimens.

> why we consider some minerals to be "gems" and how the shiny facets of gems in jewelry can be produced.

I died a mineral, and became a plant. I died as plant and rose to animal, I died as animal and I was Man. Why should I fear?

—Jalal-Uddin Rumi
(Persian mystic and poet, 1207–1273)

3.1 Introduction

Zabargad Island rises barren and brown above the Red Sea, about 70 km off the coast of southern Egypt. Nothing grows on Zabargad, except for scruffy grass and a few shrubs, so no one lives there now. But in ancient times, many workers toiled on this 5-square-km patch of desert, gradually chipping their way into the side of its highest hill. They were searching for glassy green, pea-sized pieces of peridot, a prized gem. Carefully polished peridots were worn as jewelry by ancient Egyptians. Eventually, some of the gems appeared in Europe, set into crowns and scepters. These peridots now glitter behind glass cases in museums, millennia after first being pried free from the Earth, and perhaps 10 million years after first being formed by the bonding together of still more ancient atoms.

Peridot is the gem version of olivine, one of about 4,000 minerals that have been identified on Earth so far. Mineralogists, people who specialize in the study of minerals, discover 50 to 100 new minerals every year. Each different mineral has a name. Some names come from Latin, Greek, German, or English words describing a certain characteristic; some honor a person; some indicate the place where the mineral was first recognized; and some reflect a particular element in the mineral. Some names (quartz, calcite) may be familiar, whereas others (olivine, biotite) may be less so. Although the vast majority of mineral types are rare, forming only under special conditions, many are quite common and occur in a variety of rock types at Earth's surface

Why study minerals? Without exaggeration, we can say that *minerals are the building blocks of our planet*. To a geologist, almost any study of Earth materials depends on an understanding of minerals, for minerals make up most of the rocks and sediments comprising the Earth and its landscapes. Minerals are also important from a practical standpoint (see Chapter 12). Industrial minerals serve as the raw materials for manufacturing chemicals, concrete, and wallboard. Ore minerals are the source of valuable metals like copper and gold and provide energy resources like uranium (**Fig. 3.1a, b**). And particularly beautiful forms of minerals—gems—delight the eye in jewelry. Unfortunately, though, some minerals pose environmental hazards. No wonder **mineralogy**, the study of minerals, fascinates professionals and amateurs alike.

In this chapter, we begin by presenting the geologic definition of a mineral, and look at how minerals grow. Then, we discuss the main characteristics that enable us to identify specific samples. Finally, we describe the basic scheme that mineralogists use to classify minerals. This chapter assumes that you understand the fundamental concepts of matter and energy, especially the nature of atoms, molecules, and chemical bonds. If you are rusty on these topics, please study **Box 3.1**.

3.2 What Is a Mineral?

To a geologist, a **mineral** is a naturally occurring solid, formed by geologic processes, that has a crystalline structure and a definable chemical composition. Almost all minerals are inorganic. Let's pull apart this mouthful of a definition and examine its meaning in detail.

> *Naturally occurring*: True minerals are formed in nature, not in factories. We need to emphasize this point because in recent decades, industrial chemists have learned how to synthesize materials that have characteristics virtually identical to those of real minerals. These materials are not minerals in a geologic sense, though they are referred to in the commercial world as synthetic minerals.

> *Formed by geologic processes*: Traditionally, this phrase implied processes, such as solidification of molten rock or direct precipitation from a water solution, that did not involve living organisms. Increasingly, however, geologists recognize that life is an integral part of the Earth System. So, some geologists consider solid, crystalline materials produced by organisms to be minerals too. To avoid confusion, the term "biogenic mineral" may be used when discussing such materials.

> *Solid*: A solid is a state of matter that can maintain its shape indefinitely, and thus will not conform to the shape of its container. Liquids (such as oil or water) and gases (such as air) are not minerals (see Box 3.1).

> *Crystalline structure*: The atoms that make up a mineral are not distributed randomly and cannot move around easily. Rather, they are fixed in a specific, orderly pattern. A material in which atoms are fixed in an orderly pattern is called a crystalline solid.

> *Definable chemical composition*: This simply means that it is possible to write a chemical formula for a mineral (see Box 3.1). Some minerals contain only one element, but most are compounds of two or more elements. For example, diamond and graphite have the formula C, because they consist entirely of carbon. Quartz has the formula SiO_2—it contains the elements silicon and oxygen in the proportion of one silicon atom for every two oxygen atoms. Calcite has the formula $CaCO_3$, meaning it consists of a calcium (Ca^+) ion and a carbonate (CO_3^-) ion. Some formulas are more complicated: for example, the formula for biotite is $K(Mg,Fe)_3(AlSi_3O_{10})(OH)_2$.

> *Inorganic*: Organic chemicals are molecules containing some carbon-hydrogen bonds. Sugar ($C_{12}H_{22}O_{11}$), for example,

FIGURE 3.1 Copper ore contains minerals that serve as a source of copper metal.

Malachite grows by precipitation, in a succession of layers.

(a) Malachite is a mineral contained in copper ore [$Cu_2(CO_3)(OH)_2$]; it contains copper plus other chemicals.

(b) The copper for pots is produced by processing ore minerals.

FIGURE 3.2 The nature of crystalline and noncrystalline materials.

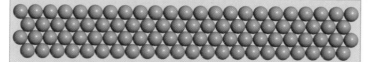

(a) This quartz crystal contains an orderly arrangement of atoms. The arrangement resembles scaffolding.

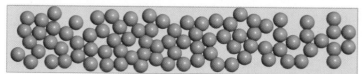

(b) Atoms in noncrystalline solids, such as glass, are not orderly.

is an organic chemical. *Almost all minerals are inorganic.* Thus, sugar and protein are not minerals. But, we have to add the qualifier "almost all" because mineralogists do consider about 30 organic substances formed by "the action of geologic processes on organic materials" to be minerals. Examples include the crystals that grow in ancient deposits of bat guano.

With these definitions in mind, we can make an important distinction between minerals and **glass**. Both minerals and glass are solids, in that they can retain their shape indefinitely. But a mineral is crystalline, and glass is not. Whereas atoms, ions, or molecules in a mineral are ordered into a crystal lattice, like soldiers standing in formation, those in a glass are arranged in a semi-chaotic way, like people at a party, in small clusters or chains that are neither oriented in the same way nor spaced at regular intervals (**Fig. 3.2a, b**).

If you ever need to figure out whether a substance is a mineral or not, just check it against the criteria listed above. Is motor oil a mineral? No—it's an organic liquid. Is table salt a mineral? Yes—it's a solid crystalline compound with the formula NaCl. Is the hard material making up the shell of an oyster considered to be a mineral? Microscopic examination of an oyster shell reveals that it consists of calcite, so it can be called a biogenic mineral. Is rock candy a mineral? No. Even though it is solid and crystalline, it's made by people and it consists of sugar (an organic chemical).

> **Did you ever wonder...**
> is rock candy a mineral?

> **Take-Home Message**
>
> Minerals are solids with a crystalline structure (an orderly arrangement of atoms inside) and a definable chemical formula. They form by natural processes in the Earth System.

3.3 Beauty in Patterns: Crystals and Their Structure

What Is a Crystal?

The word crystal brings to mind sparkling chandeliers, elegant wine goblets, and shiny jewels. But, as is the case with the word mineral, geologists have a more precise definition. A **crystal** is a single, continuous (that is, uninterrupted) piece of a crystalline solid, typically bounded by flat surfaces, called **crystal faces**, that grow naturally as the mineral forms. The word comes from the Greek *krystallos*, meaning ice. Many crystals have beautiful shapes that look like they belong in the pages of a geometry book. The angle between two adjacent crystal faces of one specimen is identical to the angle between the corresponding faces of another specimen. For example, a perfectly formed quartz crystal looks like an obelisk (**Fig. 3.3a, b**); the angle between the faces of the columnar part of a quartz crystal is always exactly 120°. This rule, discovered by one of the first geologists, Nicolas Steno (1638–1686) of Denmark, holds regardless of whether the whole crystal is big or small and regardless of whether all of the faces are the same size. Crystals come in a great variety of shapes, including cubes, trapezoids, pyramids, octahedrons, hexagonal columns, blades, needles, columns, and obelisks (**Fig. 3.3c**).

Because crystals have a regular geometric form, people have always considered them to be special, perhaps even a source of magical powers. For example, shamans of some cultures relied on talismans or amulets made of crystals, which supposedly brought power to their wearer or warded off evil spirits. Scientists have concluded, however, that crystals have no effect on health or mood. For millennia, crystals have inspired awe because of the way they sparkle, but such behavior is simply a consequence of how crystal structures interact with light.

Looking Inside Crystals

What makes crystals have regular geometric forms? This problem was the focus of study for centuries. An answer finally came from the work of a German physicist, Max von Laue, in 1912. He showed that an X-ray beam passing through a crystal breaks up into many tiny beams to create a pattern of dots on a screen (**Fig. 3.4a**). Physicists refer to this phenomenon as diffraction; it occurs when waves interact with regularly spaced objects whose spacing is close to the wavelength of the waves—you can see diffraction of ocean waves when they pass through gaps in a seawall. Von Laue concluded that, for a crystal to

BOX 3.1   CONSIDER THIS ...

Some Basic Concepts from Chemistry

To describe minerals, we need to use several terms from chemistry. To avoid confusion, terms are listed in an order that permits each successive term to utilize previous terms.

> *Element:* A pure substance that cannot be separated into other materials.

> *Atom:* The smallest piece of an element that retains the characteristics of the element. An atom consists of a nucleus surrounded by a cloud of orbiting electrons; the nucleus is made up of protons and neutrons (except in hydrogen, whose nucleus contains only one proton and no neutrons). Electrons have a negative charge, protons have a positive charge, and neutrons have a neutral charge. An atom that has the same number of electrons as protons is said to be neutral, in that it does not have an overall electrical charge.

> *Atomic number:* The number of protons in an atom of an element.

> *Atomic weight:* Approximately the number of protons plus neutrons in an atom of an element.

> *Ion:* An atom that is not neutral. An ion that has an excess negative charge (because it has more electrons than protons) is an *anion*, whereas an ion that has an excess positive charge (because it has more protons than electrons) is a *cation*. We indicate the charge with a superscript. For example, Cl^- has a single excess electron; Fe^{2+} is missing two electrons.

> *Chemical bond:* An attractive force that holds two or more atoms together (**Fig. Bx3.1a–c**). For example, *covalent bonds* form when atoms share electrons. *Ionic bonds* form when a cation and anion (ions with opposite charges) get close together and attract each other. In materials with *metallic bonds*, some of the electrons can move freely.

> *Molecule:* Two or more atoms bonded together. The atoms may be of the same element or of different elements.

> *Compound:* A pure substance that can be subdivided into two or more elements. The smallest piece of a compound that retains the characteristics of the compound is a molecule.

FIGURE Bx3.1 Examples of states of matter and chemical bonds.

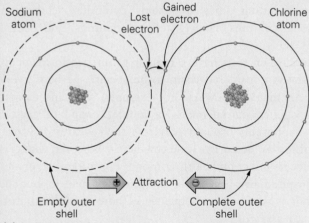

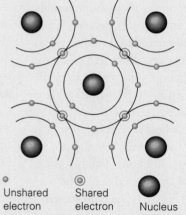

(a) An ionic bond forms between a positive ion of sodium (Na^+) and chloride (Cl^-), a negative ion of chlorine produces halite NaCl, when sodium gives up one electron to chloride, so that both have filled shells.

(b) Covalent bonds form when carbon atoms share electrons so that all have filled electron shells.

(c) In metallically bonded material, nuclei and their inner shells of electrons float in a "sea" of free electrons. The electrons stream through the metal if there is an electrical current.

cause diffraction, atoms within it must be regularly spaced and the spacing must be comparable to the wavelength of X-rays. Eventually, Von Laue and others learned how to use X-ray diffraction patterns as a basis for defining the specific arrangement of atoms in crystals. This arrangement defines the **crystal structure** of a mineral.

If you've ever examined wallpaper, you've seen an example of a pattern (**Fig. 3.4b**). Crystal structures contain one of nature's most spectacular examples of such a pattern. In crystals, the pattern is defined by the regular spacing of atoms and, if the crystal contains more than one element, by the regular alternation of atoms (**Fig. 3.4c**). (Mineralogists refer to a 3-D geometry of points representing this pattern as a lattice.) The pattern of atoms in a crystal may control the shape of a crystal. For example, if atoms in a crystal pack into the shape of a cube, the crystal may have faces that intersect at 90° angles—galena (PbS) and halite (NaCl) have such a cubic shape. Because of the pattern of atoms in a crystal structure, the structure has

> *State of matter:* The form of a substance, which reflects the degree to which the atoms or molecules comprising the matter are bonded together. **Figure Bx3.1d–f** defines three of the states—solid, liquid, and gas. There are more bonds in a solid than in a liquid, and more in a liquid than in a gas. Which state exists at a given location depends on pressure and temperature, as indicated by a phase diagram (**Fig. Bx3.1g**). A fourth state, plasma, exists only at very high temperatures.

> *Chemical:* A general name used for a pure substance (either an element or a compound).

> *Chemical formula:* A shorthand recipe that itemizes the various elements in a chemical and specifies their relative proportions. For example, the formula for water, H_2O, indicates that water consists of molecules in which two hydrogens bond to one oxygen.

> *Chemical reaction:* A process that involves the breaking or forming of chemical bonds. Chemical reactions can break molecules apart or create new molecules and/or isolated atoms.

> *Mixture:* A combination of two or more elements or compounds that can be separated without a chemical reaction. For example, a cereal composed of bran flakes and raisins is a mixture—you can separate the raisins from the flakes without destroying either.

> *Solution:* A type of material in which one chemical (the solute) dissolves in another (the solvent). In solutions, a solute may separate into ions during the process. For example, when salt (NaCl) dissolves in water, it separates into sodium (Na^+) and chloride (Cl^-) ions. In a solution, atoms or molecules of the solvent surround atoms, ions, or molecules of the solute.

> *Precipitate:* A compound that forms when ions in liquid solution join together

to create a solid that settles out of the solution; (verb) the process of forming solid grains by separation and settling from a solution. For example, when saltwater evaporates, solid salt crystals precipitate.

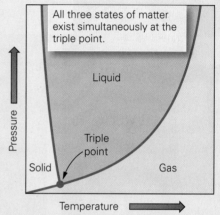

All three states of matter exist simultaneously at the triple point.

(g) The state of matter depends on pressure and temperature, as depicted in this graph, called a phase diagram.

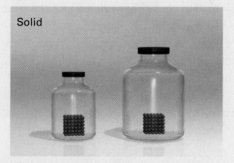

(d) A solid retains its shape regardless of the size of the container.

(e) A liquid conforms to the shape of the container, so its density does not change when the shape of the container changes.

(f) A gas expands to fill whatever volume it occupies, so its density changes if the volume changes.

symmetry, meaning that the shape of one part of the structure is the mirror image of the shape of a neighboring part. For example, if you were to cut a halite crystal or a water crystal (snowflake) in half, and place the half against a mirror, it would look whole again (**Fig. 3.4d**).

To illustrate crystal structures, we look at a few examples. Halite (rock salt) consists of oppositely charged ions that stick together because opposite charges attract. In halite, six chloride (Cl^-) ions surround each sodium (Na^+) ion, producing an overall arrangement of atoms that defines the shape of a cube (**Fig. 3.5a, b**). Diamond, by contrast, is a mineral made entirely of carbon. In diamond, each atom bonds to four neighbors arranged in the form of a tetrahedron; some naturally formed diamond crystals have the shape of a double tetrahedron (**Fig. 3.5c**). Graphite, another mineral composed entirely of carbon, behaves very differently from diamond. In contrast to diamond, graphite is so soft that we use it as the "lead" in a pencil; when a pencil moves across paper, tiny flakes of graphite peel off the pencil point and adhere to the paper. This behavior occurs because the carbon atoms in graphite are not arranged in tetrahedra, but rather occur in sheets (**Fig. 3.5d**). The sheets are bonded to each other by weak bonds and thus can separate from each other easily. Of note, two different minerals (such as diamond and graphite) that have the same composition but different crystal structures are **polymorphs**.

FIGURE 3.3 Some characteristics of crystals.

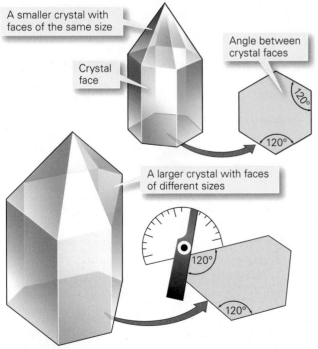

(a) A quartz crystal can resemble an obelisk. Inside, atoms are arranged in a specific geometric pattern, like the joint points in scaffolding.

(b) Regardless of specimen size, the angle between two adjacent crystal faces is consistent in a particular mineral.

The Formation and Destruction of Minerals

New mineral crystals can form in five ways. First, they can form by the solidification of a melt, meaning the freezing of a liquid to form a solid. For example, ice crystals, a type of mineral, are made by solidifying water, and many different minerals form by solidifying molten rock. Second, they can form by precipitation from a solution, meaning that atoms, molecules, or ions dissolved in water bond together and separate out of the water. Salt crystals, for example, precipitate when you evaporate salt water (see Box 3.1). Third, they can form by solid-state diffusion, the movement of atoms or ions through a solid to arrange into a new crystal structure, a process that takes place very slowly. For example, garnets grow by diffusion in solid rock. Fourth, minerals can form at interfaces between the physical and biological components of the Earth System by a process called biomineralization. This occurs when living organisms cause minerals to precipitate either within or on their bodies, or immediately adjacent to their bodies. For example, clams and other shelled organisms extract ions from water to produce mineral shells. Fifth, minerals can precipitate directly from a gas. This process typically occurs around volcanic vents or around geysers, for at such locations volcanic gases or steam enter the atmosphere and cool abruptly. Some of the bright yellow sulfur deposits found in volcanic regions form in this way.

The first step in forming a crystal is the chance formation of a seed, or an extremely small crystal (**Fig. 3.6a**). Once the seed exists, other atoms in the surrounding material attach themselves to the face of the seed. As the crystal grows, crystal faces move outward but maintain the same orientation (**Fig. 3.6b**). The youngest part of the crystal is at its outer edge.

In the case of crystals formed by the solidification of a melt, atoms begin to attach to the seed when the melt becomes so cool that thermal vibrations can no longer break apart the

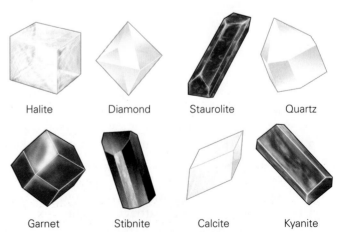

(c) Crystals come in a variety of shapes, including cubes, prisms, blades, and pyramids.

FIGURE 3.4 Patterns and symmetry in minerals.

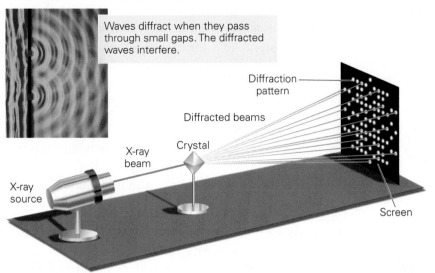

Waves diffract when they pass through small gaps. The diffracted waves interfere.

Diffraction pattern

Diffracted beams

Crystal

X-ray beam

X-ray source

Screen

(a) Diffraction of an X-ray beam passing through a crystal produces a pattern of bright spots on a screen. The spots are due to interference of overlapping light waves.

Sulfur
Lead

(b) The repetition of a flower motif on wallpaper.

(c) The repetition of alternating sulfur and lead atoms in the mineral galena (PbS).

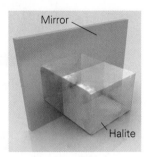

Mirror

Halite

Mirror

Snowflake

(d) Minerals display symmetry. One-half of a crystal is a mirror image of the other.

attraction between the seed and the atoms in the melt. Crystals formed by precipitation from a solution develop when the solution becomes saturated, meaning the number of dissolved ions per unit volume of solution becomes so great that they can get close enough to each other to bond together.

As crystals grow, they develop their particular crystal shape, based on the geometry of their internal structure. The shape is defined by the relative dimensions of the crystal (needle-like, sheet-like, etc.) and the angles between crystal faces. Typically, the growth of minerals is restricted in one or more directions, because existing crystals act as obstacles. In such cases, minerals grow to fill the space that is available, and their shape is controlled by the shape of their surroundings. Minerals without well-formed crystal faces are anhedral grains (**Fig. 3.6c**). If a mineral's growth is unimpeded so that it displays well-formed crystal faces, then it is a euhedral crystal. The surface crystals of a **geode**, a mineral-lined cavity in rock, may be euhedral (**Fig. 3.6d**).

A mineral can be destroyed by melting, dissolving, or some other chemical reaction. Melting involves heating a mineral to a temperature at which thermal vibration of the atoms or ions in the lattice break the chemical bonds holding them to the lattice. The atoms or ions then separate, either individually or in small groups, to move around again freely. Dissolution occurs when you immerse a mineral in a solvent, such as water. Atoms or ions then separate from the crystal face and are surrounded by solvent molecules. Chemical reactions can destroy a mineral when it comes in contact with reactive materials. For example, iron-bearing minerals react with air and water to form rust. The action of microbes in the environment can also destroy minerals. In effect, some microbes can "eat" certain minerals; the microbes use the energy stored in the

chemical bonds that hold the atoms of the mineral together as their source of energy for metabolism.

> **Take-Home Message**
>
> The crystal structure of minerals is defined by a regular geometric arrangement of atoms that has symmetry. Minerals can form by solidification of a melt, by precipitation from a water solution or a gas, or by rearrangement of atoms in a solid.

3.4 How Can You Tell One Mineral From Another?

Amateur and professional mineralogists get a kick out of recognizing minerals. They might hover around a display case in a museum and name specimens without bothering to look at the labels. How do they do it? The trick lies in learning to recognize the basic physical properties (visual and material characteristics) that distinguish one mineral from another. Some physical properties, such as shape and color, can be seen from a distance. Others, such as hardness and magnetization, can be determined only by handling the specimen or by performing an identification test on it. Identification tests include scratching the mineral by another object, placing it near a magnet, weighing it, tasting it, or placing a drop of acid on it. Let's examine some of the physical properties most commonly used in basic mineral identification.

FIGURE 3.5 The nature of crystalline structure in minerals. The arrangement of atoms can be portrayed by a ball-and-stick model, or by a packed-ball model.

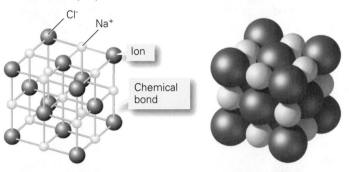

(a) In a ball-and-stick model of halite, the balls are ions, and the sticks are chemical bonds.

(b) This packed-ball model gives a better sense of how ions fit together in crystal.

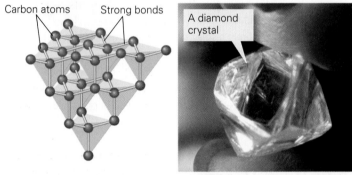

(c) In a diamond, carbon atoms are arranged in tetrahedra. All of the bonds are strong.

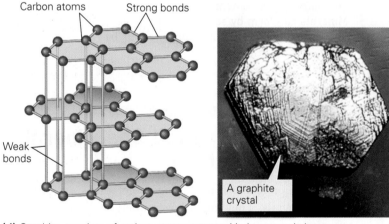

(d) Graphite consists of carbon atoms arranged in hexagonal sheets. The sheets are connected by weak bonds.

> *Color*: **Color** results from the way a mineral interacts with light. Sunlight contains the whole spectrum of colors; each color has a different wavelength. A mineral absorbs certain wavelengths, so the color you see when looking at a specimen represents the wavelengths the mineral does not absorb. Certain minerals always have the same color, but many dis-

play a range of colors (**Fig. 3.7a**). Color variations in a mineral are due to the presence of impurities. For example, trace amounts of iron may give quartz a reddish color.

> *Streak*: The **streak** of a mineral refers to the color of a powder produced by pulverizing the mineral. You can obtain a streak by scraping the mineral against an unglazed ceramic plate (**Fig. 3.7b**). The color of a mineral powder tends to be less variable than the color of a whole crystal, and thus provides a fairly reliable clue to a mineral's identity. Calcite, for example, always yields a white streak even though pieces of calcite may be white, pink, or clear.

> *Luster*: **Luster** refers to the way a mineral surface scatters light. Geoscientists describe luster by comparing the appearance of the mineral with the appearance of a familiar substance. For example, minerals that look like metal have a metallic luster, whereas those that do not have a nonmetallic luster—the adjectives are self-explanatory (**Fig. 3.7c, d**). Terms used for types of nonmetallic luster include silky, glassy, satiny, resinous, pearly, or earthy.

> *Hardness*: **Hardness** is a measure of the relative ability of a mineral to resist scratching, and it therefore represents the resistance of bonds in the crystal structure to being broken. The atoms or ions in crystals of a hard mineral are more strongly bonded than those in a soft mineral. Hard minerals can scratch soft minerals, but soft minerals cannot scratch hard ones. Diamond, the hardest mineral known, can scratch most anything, which is why it is used to cut glass. In the early 1800s, a mineralogist named Friedrich Mohs listed some minerals in sequence of relative hardness; a mineral with a hardness of 5 can scratch all minerals with a hardness of 5 or less. This list, the **Mohs hardness scale**, helps in mineral identification. To make the scale easy to use, common items such as your fingernail, a penny, or a glass plate have been added (**Table. 3.1**).

> *Specific gravity*: **Specific gravity** represents the density of a mineral, as represented by the ratio between the weight of a volume of the mineral and the weight of an equal volume of water at 4°C. For example, one cubic centimeter of quartz has a weight of 2.65 grams, whereas one cubic centimeter of water has a weight of 1.00 gram. Thus, the specific gravity of quartz is 2.65. In practice, you can develop a "feel" for specific gravity by hefting minerals in your hands. A piece of galena (lead ore) feels heavier than a similar-sized piece of quartz.

> *Crystal habit*: The **crystal habit** of a mineral refers to the shape of a single crystal with well-formed crystal faces, or to the character of an aggregate of many well-formed crystals that grew together as a group (**Fig. 3.7e**). The habit depends on the internal arrangement of atoms in the crystal. A description of habit generally includes adjectives that

FIGURE 3.6 The growth of crystals.

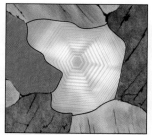

Ions attach to
the crystal face.

(a) New crystals nucleate and begin to precipitate out of a water solution. As time progresses, they grow into the open space.

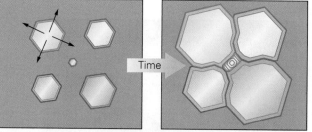

(b) New crystals grow outward from the central seed. As time passes, they maintain their shape until they interfere with each other.

(c) A crystal growing in a confined space will be anhedral.

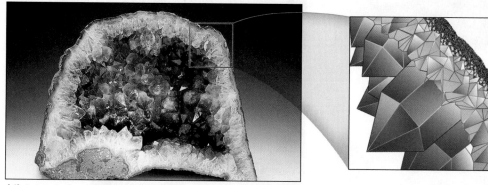

(d) A geode from Brazil consists of purple quartz crystals (amethyst) that grew from the wall into the center. The enlargement sketch indicates that the crystals are euhedral.

highlight the shape of the crystal. For example, crystals that are roughly the same length in all directions are called equant or blocky, those that are much longer in one dimension than in others are columnar or needle-like, those shaped like sheets of paper are platy, and those shaped like knives are bladed.

> *Special properties*: Some minerals have distinctive properties that readily distinguish them from other minerals. For example, calcite ($CaCO_3$) reacts with dilute hydrochloric acid (HCl) to produce carbon dioxide (CO_2) gas (**Fig. 3.7f**). Dolomite ($CaMg[CO_3]_2$) also reacts with acid, but not as strongly. Graphite makes a gray mark on paper, magnetite attracts a magnet (**Fig. 3.7g**), halite tastes salty, and plagioclase has striations (thin parallel corrugations or stripes) on its surface.

> *Fracture and cleavage*: Different minerals fracture (break) in different ways, depending on the internal arrangement of atoms. If a mineral breaks to form distinct planar surfaces that have a specific orientation in relation to the crystal structure, then we say that the mineral has **cleavage** and we refer to each surface as a cleavage plane. Cleavage forms in directions where the bonds holding atoms together in the crystal are the weakest (**Fig. 3.8a–e**). Some minerals have one direction of cleavage. For example, mica has very weak bonds in one direction but strong bonds in the other two directions. Thus, it easily splits into parallel sheets; the surface of each sheet is a cleavage plane. Other minerals have two or three directions of cleavage that intersect at a specific angle. For example, halite has three sets of cleavage planes that intersect at right angles, so halite crystals break into little cubes. Materials that have no cleavage at all (because bonding is equally strong in all directions) break either by forming irregular fractures or by forming conchoidal fractures (**Fig. 3.8f**). **Conchoidal fractures** are smoothly curving, clamshell-shaped surfaces; they typically form in glass. Cleavage planes are sometimes hard to distinguish from crystal faces (**Fig. 3.8g**).

Take-Home Message

The characteristics of minerals (such as color, streak, luster, crystal shape, hardness, specific gravity, cleavage, magnetism, and reaction with acid) are a manifestation of the crystal structure and chemical composition of minerals.

FIGURE 3.7 Physical characteristics of minerals.

(a) Color is diagnostic of some minerals, but not all. For example, quartz can come in many colors.

(b) To obtain the streak of a mineral, rub it against a porcelain plate. The streak consists of mineral powder.

(c) Pyrite has a metallic luster because it gleams like metal.

(d) Feldspar has a nonmetallic luster.

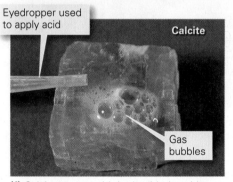

(f) Calcite reacts with hydrochloric acid to produce carbon dioxide gas.

(e) Crystal habit refers to the shape or character of the crystal. The blue kyanite crystals on the left are bladed, and the chrysotile on the right is fibrous.

(g) Magnetite is magnetic.

TABLE 3.1 Mohs hardness scale. Mohs' numbers are relative—in reality, diamond is 3.5 times harder than corundum, as the graph shows.

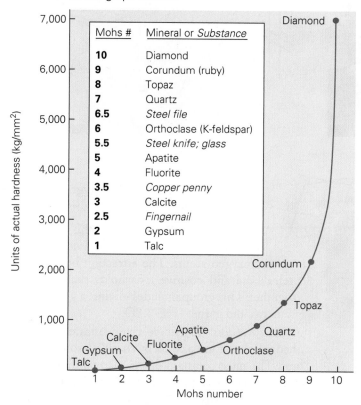

Mohs #	Mineral or *Substance*
10	Diamond
9	Corundum (ruby)
8	Topaz
7	Quartz
6.5	*Steel file*
6	Orthoclase (K-feldspar)
5.5	*Steel knife; glass*
5	Apatite
4	Fluorite
3.5	*Copper penny*
3	Calcite
2.5	*Fingernail*
2	Gypsum
1	Talc

3.5 Organizing Our Knowledge: Mineral Classification

The 4,000 known minerals can be separated into a small number of groups, or mineral classes. You may think, "Why bother?" Classification schemes are useful because they help organize information and streamline discussion. Biologists, for example, classify animals into groups based on how they feed their young and on the architecture of their skeletons, and botanists classify plants according to the way they reproduce and by the shape of their leaves. In the case of minerals, a good means of classification eluded researchers until it became possible to determine the chemical makeup of minerals. A Swedish chemist, Baron Jöns Jacob Berzelius (1779–1848), analyzed minerals and noted chemical similarities among many of them. Berzelius, along with his students, established that most minerals can be classified by specifying the principal anion (negative ion) or anionic group (negative molecule) within the mineral (see Box 3.1). We now take a look at principal mineral classes, focusing especially on silicates, the class that constitutes most of the rock in the Earth.

The Mineral Classes

Mineralogists distinguish several principal classes of minerals. Here are some of the major ones.

> *Silicates*: The fundamental component of most silicates in the Earth's crust is the SiO_4^{4-} anionic group. A well-known example, quartz (Fig. 3.7a), has the formula SiO_2. We will learn more about silicates in the next section.

> *Oxides*: Oxides consist of metal cations bonded to oxygen anions. Typical oxide minerals include hematite (Fe_2O_3; Fig. 3.7b) and magnetite (Fe_3O_4; Fig. 3.7g).

> *Sulfides*: Sulfides consist of a metal cation bonded to a sulfide anion (S^{2-}). Examples include galena (PbS) and pyrite (FeS_2; Fig. 3.7c).

> *Sulfates*: Sulfates consist of a metal cation bonded to the SO_4^{2-} anionic group. Many sulfates form by precipitation out of water at or near the Earth's surface. An example is gypsum ($CaSO_4 \cdot 2H_2O$).

> *Halides*: The anion in a halide is a halogen ion (such as chloride [Cl^-] or fluoride [F^-]), an element from the second column from the right in the periodic table (see Appendix). Halite, or rock salt (NaCl; Fig. 3.8d), and fluorite (CaF_2), a source of fluoride, are common examples.

> *Carbonates*: In **carbonates**, the molecule CO_3^{2-} serves as the anionic group. Elements such as calcium or magnesium bond to this group. The two most common carbonates are calcite ($CaCO_3$; Fig. 3.8e) and dolomite ($CaMg[CO_3]_2$).

> *Native metals*: Native metals consist of pure masses of a single metal. The metal atoms are bonded by metallic bonds (see Box 3.1). Copper and gold, for example, may occur as native metals.

Silicates: The Major Rock-Forming Minerals

Silicate minerals, or **silicates**, make up over 95% of the continental crust and almost 100% of the oceanic crust and of the Earth's mantle consist almost entirely of silicates. Thus, silicates are the most common minerals on Earth. As we've noted, silicates in the Earth's crust and upper mantle contain the SiO_4^{4-} anionic group. In this group, four oxygen atoms surround a single silicon atom, thereby defining the corners of a tetrahedron, a pyramid-like shape with four triangular faces (**Fig. 3.9a**). We refer to this anionic group as the **silicon-oxygen tetrahedron** (or, informally, as the silica tetrahedron), and it acts, in effect, as the building block of silicate minerals.

Mineralogists distinguish among several groups of silicate minerals based on the way in which silica tetrahedra are arranged (**Fig. 3.9b**). The arrangement, in turn, determines the degree to which tetrahedra share oxygen atoms. Note that the number of shared oxygens determines the ratio of silicon (Si) to oxygen (O) in the mineral. Here are the groups, in

FIGURE 3.8 The nature of mineral cleavage and fracture.

(a) Mica has one strong plane of cleavage and splits into sheets.

(b) Pyroxene has two planes of cleavage that intersect at 90°.

(c) Amphibole has two planes that intersect at 60°.

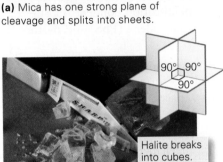

Halite breaks into cubes.

(d) Halite has three mutually perpendicular planes of cleavage.

Calcite breaks into rhombs.

(e) Calcite has three planes of cleavage, none of which are perpendicular to the others.

Irregular fracture

Crystal face

Garnet

Conchoidal fracture

Quartz

(f) Minerals without cleavage can develop irregular or conchoidal fractures.

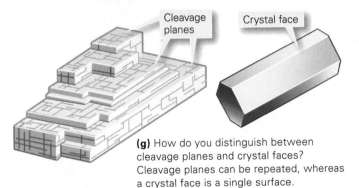

Cleavage planes

Crystal face

(g) How do you distinguish between cleavage planes and crystal faces? Cleavage planes can be repeated, whereas a crystal face is a single surface.

order from fewer shared oxygens to more shared oxygens:

> *Independent tetrahedra*: In this group, the tetrahedra are independent and do not share any oxygen atoms. The attraction between the tetrahedra and positive ions holds such minerals together. This group includes olivine, a glassy green mineral, and garnet (Fig. 3.8f).

> *Single chains*: In a single-chain silicate, the tetrahedra link to form a chain by sharing two oxygen atoms. The most common of the many different types of single-chain silicates are pyroxenes (Fig. 3.8b).

> *Double chains*: In a double-chain silicate, the tetrahedra link to form a double chain by sharing two or three oxygen atoms. Amphiboles are the most common type (Fig. 3.8c).

> *Sheet silicates*: The tetrahedra in this group share three oxygen atoms and therefore link to form two-dimensional sheets. Other ions and, in some cases, water molecules fit between the sheets in some sheet silicates. Because of their structure, sheet silicates have cleavage in one direction, and they occur in books of very thin sheets. In this group, we find micas (Fig. 3.8a) and clays. Clays occur only in extremely tiny flakes.

> *Framework silicates*: In a framework silicate, each tetrahedron shares all four oxygen atoms with its neighbors, forming a three-dimensional structure. Examples include feldspar and quartz. The two most common feldspars are plagioclase, which tends to be white, gray, or blue; and orthoclase (also called potassium feldspar, or K-feldspar), which tends to be pink (Fig. 3.7d).

Take-Home Message

The 4,000 known minerals can be organized into a relatively small number of classes based on chemical makeup. Most minerals are silicates, which contain silicon-oxygen tetrahedral arranged in various ways.

FIGURE 3.9 The structure of silicate minerals.

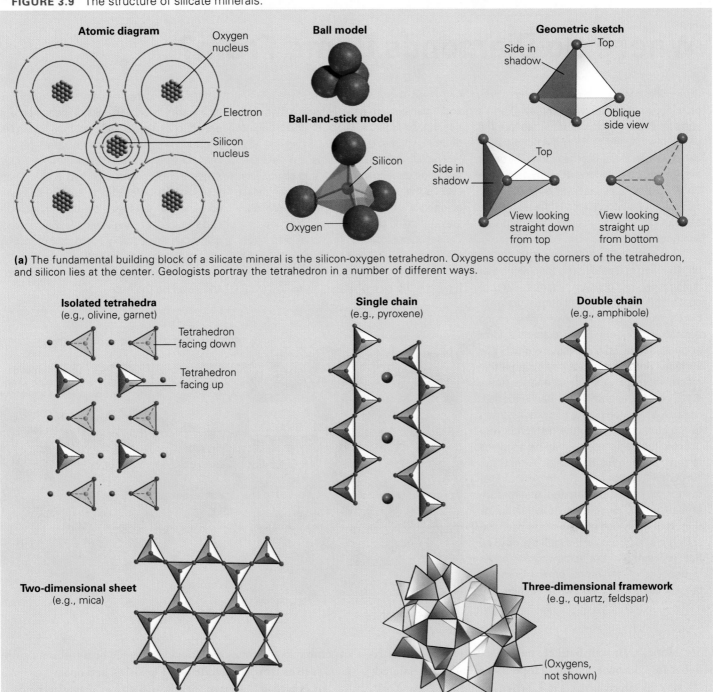

Atomic diagram

Oxygen nucleus

Electron

Silicon nucleus

Ball model

Ball-and-stick model

Silicon

Oxygen

Geometric sketch

Top

Side in shadow

Oblique side view

Top

Side in shadow

View looking straight down from top

View looking straight up from bottom

(a) The fundamental building block of a silicate mineral is the silicon-oxygen tetrahedron. Oxygens occupy the corners of the tetrahedron, and silicon lies at the center. Geologists portray the tetrahedron in a number of different ways.

Isolated tetrahedra
(e.g., olivine, garnet)

Tetrahedron facing down

Tetrahedron facing up

Single chain
(e.g., pyroxene)

Double chain
(e.g., amphibole)

Two-dimensional sheet
(e.g., mica)

Three-dimensional framework
(e.g., quartz, feldspar)

(Oxygens, not shown)

(b) The classes of silicate minerals differ from one another by the way in which the silicon-oxygen tetrahedra are linked. Where the tetrahedra link, they share an oxygen atom. Oxygen atoms are shown in red. Positive ions (not shown) occupy spaces between tetrahedra.

3.6 Something Precious—Gems!

Mystery and romance follow famous gems. Consider the stone now known as the Hope Diamond, recognized by name the world over (**Fig. 3.10**). No one knows who first dug it out of the ground (**Box 3.2**). Was it mined in the 1600s, or was it stolen off an ancient religious monument? What we do know is that in the 1600s, a French trader named Jean Baptiste Tavernier obtained a large (112.5 carats, where 1 carat = 200 milligrams), rare blue diamond in India, perhaps from a Hindu statue, and carried it back to France. King Louis XIV bought the diamond and had it fashioned into a jewel of 68 carats. This jewel vanished in 1762 during a burglary. Perhaps it was lost forever—perhaps not. In 1830, a 44.5-carat blue diamond mysteriously appeared on the jewel market for sale.

> **Did you ever wonder...**
> where diamonds come from and how they form?

BOX 3.2   CONSIDER THIS ...

Where Do Diamonds Come From?

Diamonds consist of carbon, which typically accumulates only at or near Earth's surface. Experiments demonstrate that the pressures needed to form diamond are so extreme that, in nature, they generally occur only at depths of around 150 km below the Earth's surface. Nowadays, engineers can duplicate these conditions in the laboratory, so corporations manufacture several tons of synthetic diamonds a year.

How does carbon get down to depths of 150 km? Geologists speculate that subduction or collision carries carbon-containing rocks and sediments down to the depth where it transforms into diamond beneath continents. But if diamonds form at great depth, how do they return to the surface? Some diamonds rise when rifting cracks the continental crust and causes a small part of the underlying mantle to melt. Magma generated during this process rises to the surface, bringing the diamonds with it. Near the surface, the magma solidifies to form an igneous rock called kimberlite, named for Kimberley, South Africa. Diamonds brought up with the magma are embedded as crystals in solid kimberlite (**Fig. Bx3.2**). Much of the world's diamond supply comes from mines in this rock (**See For Yourself C**). But some sources occur in deposits of sediment formed from the breakdown and erosion of kimberlite that had been exposed at the surface. Rivers and glaciers may transport diamond-bearing sediments far from their original bedrock source.

Not all natural diamonds are valuable; the value depends on color and clarity. Diamonds that contain imperfections (cracks, or specks of other material), or are dark gray in color, are not used for jewelry. These stones, called industrial diamonds, are used as abrasives. Gem-quality diamonds come in a range of sizes. Jewelers measure the size of these gems in carats, where one carat equals 200 milligrams (0.2 gram). In English units of measurement, one ounce equals 142 carats. The largest diamond ever found, a stone called the Cullinan Diamond, was discovered in South Africa in 1905, and weighed 3,106 carats (621 grams) before being cut. By comparison, the diamond on a typical engagement ring weighs less than one carat. Gem-quality diamonds are actually more common than you might expect—suppliers stockpile the stones in order to avoid flooding the market and lowering the price.

FIGURE Bx3.2 Diamond occurrences.

A diamond mine pit.

A diamond embedded in solid kimberlite.

Henry Hope, a British banker, purchased the stone, which then became known as the Hope Diamond. It changed hands several times until 1958, when a famous New York jeweler named Harry Winston donated it to the Smithsonian Institution in Washington, DC, where it now sits behind bulletproof glass in a heavily guarded display.

What makes stones such as the Hope Diamond so special that people risk life and fortune to obtain them? What is the difference between a gemstone, a gem, and any other mineral? A gemstone is a mineral that has special value because it is rare and people consider it beautiful. A **gem**, or jewel, is a finished stone ready to be set in jewelry. Jewelers distinguish between precious stones (such as diamond, ruby, sapphire, and emerald), which are particularly rare and expensive, and semiprecious stones (such as topaz, tourmaline, aquamarine, and garnet), which are less rare and less expensive. All the stones mentioned so far are transparent crystals, though most have some color. The category of semiprecious stones also includes opaque or translucent minerals such as lapis, malachite (see Fig. 3.1a), and opal.

In everyday language, pearls and amber may also be considered gemstones. Unlike diamonds and garnets, which form inorganically in rocks, pearls form in living oysters when the oyster extracts calcium and carbonate ions from water and precipitates them around an impurity, such as a sand grain, embedded in its body. Thus, pearls are a result of biomineralization. Most pearls used in jewelry today are "cultured" pearls, made by artificially introducing round sand grains into oysters in order to stimulate pearl production. Amber is also formed by organic processes—it consists of fossilized tree sap. But because amber consists of organic compounds that are not arranged in a crystal structure, it does not meet the definition of a mineral.

In some cases, gemstones are merely pretty and rare versions of more common minerals. For example, ruby is a special version of the common mineral corundum, and emerald is a special version of the common mineral beryl (**Fig. 3.11a**). As for the beauty of a gemstone, this quality lies basically in its color and, in the case of transparent gems, its "fire"—the way the mineral bends and internally reflects the light passing through it, and disperses the light into a spectrum. Fire makes a diamond sparkle more than a similarly cut piece of glass.

Gemstones form in many ways. Some solidify from a melt, some form by diffusion, some precipitate out of a water solution in cracks, and some are a consequence of the chemical interaction of rock with water near the Earth's surface. Many gems come from pegmatites, particularly coarse-grained rocks formed by the solidification of steamy melt.

Most gems used in jewelry are "cut" stones, meaning that they are not raw crystals right from the ground, but rather have been faceted. The smooth **facets** on a gem are ground and polished surfaces made with a faceting machine (**Fig. 3.11b**). Facets are not the natural crystal faces of the mineral, nor are they cleavage planes, though gem cutters sometimes make the facets parallel to cleavage directions and will try to break a large gemstone into smaller pieces by splitting it on a cleavage plane. A faceting machine consists of a doping arm, a device that holds a stone in a specific orientation, and a lap, a rotating disk covered with a wet paste of grinding powder and water. The gem cutter fixes a gemstone to the end of the doping arm and positions the arm so that it holds the stone against the moving lap. The movement of the lap grinds a facet. When the facet is complete, the gem cutter rotates the arm by a specific angle, lowers the stone, and grinds another facet. The geometry of the facets defines the cut of the stone. Different cuts have names, such as "brilliant," "French," "star," and "pear." Grinding facets is a lot of work—a typical engagement-ring diamond with a brilliant cut has 57 facets (**Fig. 3.11c**)!

> **Did you ever wonder...**
> how jewelers make the facets on a jewel?

FIGURE 3.10 The Hope Diamond.

FIGURE 3.11 Cutting gemstones.

Non-gem-quality beryl

Rough emerald

Cut emerald

(a) Emerald is a green, transparent variety of the mineral beryl.

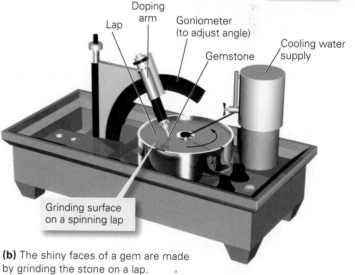

Doping arm

Lap

Goniometer (to adjust angle)

Gemstone

Cooling water supply

Grinding surface on a spinning lap

(b) The shiny faces of a gem are made by grinding the stone on a lap.

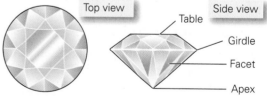

Top view

Side view

Table

Girdle

Facet

Apex

(c) There are many different "cuts" for a gem. Here we see the top and side views of a brilliant-cut diamond.

Take-Home **Message**

Gemstones are particularly rare and beautiful minerals. The gems or jewels found in jewelry have been faceted using a lap—the facets are not natural crystal faces or cleavage surfaces. The fire of a jewel comes from the way it reflects light internally.

Chapter Summary

> Minerals are naturally occurring, solid substances, formed by geologic processes, with a definable chemical composition and an internal structure characterized by an orderly arrangement of atoms, ions, or molecules in a crystalline lattice. Most minerals are inorganic.

> Biogenic minerals are produced by organisms. The minerals in shells are an example of biomineralization.

> In the crystalline lattice of minerals, atoms occur in a specific pattern—one of nature's finest examples of ordering.

> Minerals can form by the solidification of a melt, precipitation from a water solution, diffusion through a solid, the metabolism of organisms, and precipitation from a gas.

> About 4,000 different types of minerals are known, each with a name and distinctive physical properties (such as color, streak, luster, hardness, specific gravity, crystal habit, cleavage, magnetism, and reactivity with acid).

> The unique physical properties of a mineral reflect its chemical composition and crystal structure. By observing these physical properties, you can identify minerals.

> The most convenient way to classify minerals is to group them according to their chemical composition. Mineral classes include silicates, oxides, sulfides, sulfates, halides, carbonates, and native metals.

> Silicate minerals are the most common minerals on Earth. The silicon-oxygen tetrahedron, a silicon atom surrounded by four oxygen atoms, serves as the fundamental building block of silicate minerals.

> Groups of silicate minerals are distinguished from each other by the ways in which the silicon-oxygen tetrahedra that constitute them are linked.

> Gemstones are minerals known for their beauty and rarity. The facets on cut gems used in jewelry are made by grinding and polishing the stones with a faceting machine.

Key Terms

carbonate (p. 81)
cleavage (p. 79)
color (p. 78)
conchoidal fracture (p. 79)
crystal (p. 73)
crystal face (p. 73)

crystal habit (p. 78)
crystal structure (p. 74)
facet (p. 85)
gem (p. 84)
geode (p. 77)
glass (p. 73)

hardness (p. 78)
luster (p. 78)
mineral (p. 72)
mineralogy (p. 72)
Mohs hardness scale (p. 78)

polymorph (p. 76)
silicate (p. 81)
silicon-oxygen tetrahedron (p. 81)
specific gravity (p. 78)
streak (p. 78)

Review Questions

1. What is a mineral, as geologists understand the term? How is this definition different from the everyday usage of the word?

2. Why is glass not a mineral?

3. Salt is a mineral, but the plastic making up an inexpensive pen is not. Why not?

4. Describe several ways that mineral crystals can form.

5. Why do some minerals occur as euhedral crystals, whereas others occur as anhedral grains?

6. List and define the principal physical properties used to identify a mineral. Which minerals react with acid to produce CO_2?

7. How can you determine the hardness of a mineral? What is the Mohs hardness scale?

8. How do you distinguish cleavage surfaces from crystal faces on a mineral? How does each type of surface form?

9. What is the prime characteristic that geologists use to separate minerals into classes?

10. What is a silicon-oxygen tetrahedron? What is the anionic group that occurs in carbonate minerals?

11. On what basis do mineralogists organize silicate minerals into distinct groups?

12. What is the relationship between the way in which silicon-oxygen tetrahedra bond in micas and the characteristic cleavage of micas?

13. Why are some minerals considered gemstones? How do you make the facets on a gem?

Every chapter of SmartWork contains active learning exercises to assist you with reading comprehension and concept mastery. This chapter also features:

> A What a Geologist Sees exercise on identifying mineral properties.

> An Animation exercise on mineral growth.

> Problems that help students with mineral classification.

On Further Thought

14. Compare the chemical formula of magnetite with that of biotite. Considering that iron is a relatively heavy element, which mineral has the greater specific gravity?

15. Imagine that you are given two milky white crystals, each about 2 cm across. You are told that one of the crystals is composed of plagioclase and the other of quartz. How can you determine which is which?

16. Could you use crushed calcite to grind facets on a diamond? Why or why not?

SEE FOR YOURSELF C... Minerals

Download *Google Earth*™ from the Web in order to visit the locations described below (instructions appear in the Preface of this book). You'll find further locations and associated active-learning exercises on Worksheet C of our **Geotours Workbook**.

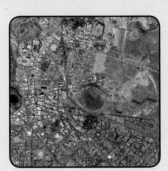

Kimberley Diamond Mine
Latitude   28°44'17.06"S,
Longitude   24°46'30.77"E

The field of view shows part of the town of Kimberley, in South Africa. Looking down from 13 km, you can see an inactive diamond mine, which looks like a circular pit, and the tailings pile of excavated rock debris.

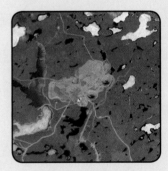

Ekati Diamond Mine, Canada
Latitude   64°43'14.74"N,
Longitude   110°37'32.76"W

In this remote region, the landscape is largely untouched tundra. In the 1990s, prospectors found diamond pipes, and now the area contains small open-pit mines. Mining is difficult on the frozen ground.

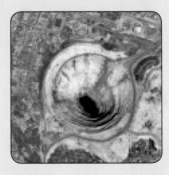

Mir Mine, Siberia
Latitude   62°31'40.77"N,
Longitude   113°59'36.16"E

Viewed from 4.2 km, we see an open pit (1.2 km across and 525 m deep) dug into kimberlite. In the 1960s, the mine produced 2,000 kg of diamonds per year. Mining continued underground after the pit closed in 2001.

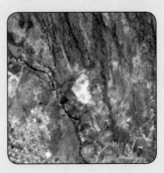

Diamantina, Brazil
Latitude   18°15'4.35"S,
Longitude   43°34'57.21"W

Independent miners excavated Precambrian sedimentary rocks by hand in this pit, viewed from 2.2 km. Diamonds occur as grains. They weathered out of kimberlite, then mixed in with sand and pebbles carried by rivers. The sediment later cemented into rock.

Rock Groups

The rock traversed by this highway in Utah tells us a story of the Earth's past.

A.1 Introduction

During the 1849 gold rush in the Sierra Nevada of California, only a few lucky individuals actually became rich. The rest of the "forty-niners" either slunk home in debt or took up less glamorous jobs in new towns such as San Francisco. These towns grew rapidly, and soon people from the American west coast were demanding large quantities of manufactured goods from east-coast factories. Making the goods was no problem, but getting them to California meant either a stormy voyage around the southern tip of South America or a trek with stubborn mule teams through the deserts of Nevada and Utah. The time was ripe to build a railroad linking the east and west coasts of North America, and, with much fanfare, the Central Pacific line decided to punch one right across the peaks of the Sierras. In 1863, while the Civil War raged elsewhere in the United States, the company transported six thousand Chinese laborers across the Pacific in the squalor of unventilated cargo holds and set them to work chipping ledges and blasting tunnels. Along the way, untold numbers of laborers died of frostbite, exhaustion, mistimed blasts, landslides, or avalanches.

Through their efforts, the railroad laborers certainly gained an intimate knowledge of how rock feels and behaves—it's solid, heavy, and hard! They also found that some rocks break easily into layers but others do not, and that some rocks are dark-colored while others are light-colored. They realized, like anyone who looks closely at rock exposures, that rocks are not just gray, featureless masses, but rather come in a great variety of colors and textures.

Why are there so many distinct types of rocks? The answer is simple: rocks can form in many different ways and from many different materials. Because of the relationship between rock type and the process of formation, rocks provide a historical record of geologic events and give insight into interactions among components of the Earth System. The next few chapters are devoted to a discussion of rocks and a description of how rocks form. This interlude serves as a general introduction to these chapters. We learn what the term "rock" means to geologists, what rocks are made of, and how to distinguish among the three principal groups of rocks. We also look at how geologists study rocks.

A.2 What Is Rock?

To geologists, **rock** is a coherent, naturally occurring solid, consisting of an aggregate of minerals or, less commonly, of glass. Let's take this definition apart to see what its components mean.

> *Coherent*: A rock holds together, and thus must be broken to be separated into pieces. As a result of its coherence, rock can form

cliffs or can be carved into sculptures. A pile of unattached mineral grains does not constitute a rock.

> *Naturally occurring*: Geologists consider only naturally occurring materials to be rocks, so manufactured materials, such as concrete and brick, do not qualify.

> *An aggregate of minerals or a mass of glass*: The vast majority of rocks consist of an aggregate (a collection) of many mineral grains, and/or crystals, stuck or grown together. Some rocks contain only one kind of mineral, whereas others contain several different kinds. A few rock types consist of glass.

What holds rock together? Grains in rock stick together to form a coherent mass either because they are bonded by natural **cement**, mineral material that precipitates from water and fills the space between grains (**Fig. A.1a**), or because they interlock with one another like pieces in a jigsaw puzzle (**Fig. A.1b**). Rocks whose grains are stuck together by cement are called **clastic**, whereas rocks whose crystals interlock with one another are called **crystalline**. Glassy rocks hold together because they originate as a continuous mass (that is, they have no separate grains), because glassy grains were welded together while still hot, or because they were cemented together at a later time.

At the surface of the Earth, rock occurs either as broken chunks (pebbles, cobbles, or boulders; see Chapter 6) that have moved by falling down a slope or by being transported in ice, water, or wind, or as **bedrock** that is still attached to the Earth's crust. Geologists refer to an exposure of bedrock as an **outcrop**. An outcrop may appear as a rounded knob out in a field, as a ledge forming a cliff or ridge, on the face of a stream cut (where running water dug down into bedrock), or along human-made roadcuts and excavations (**Fig. A.2a–d**).

To people who live in cities or forests or on farmland, outcrops of bedrock may be unfamiliar, since bedrock may be completely covered by vegetation, sand, mud, gravel, soil, water, asphalt, concrete, or buildings. Outcrops are particularly rare in regions such as the midwestern United States, where, during the past million years, ice-age glaciers melted and buried bedrock under thick deposits of debris (see Chapter 18).

A.3 The Basis of Rock Classification

Beginning in the 18th century, geologists struggled to develop a sensible way to classify rocks, for they realized, as did miners from centuries past, that not all rocks are the same. Classification schemes help us organize information and remember significant details about materials or objects, and they help us recognize similarities and differences among them. By the end of the 18th century, most geologists had accepted the *genetic scheme* for classifying rocks that we continue to use today. This scheme focuses on the origin (genesis) of rocks. Using this

FIGURE A.1   Rocks, aggregates of mineral grains and/or crystals, can be clastic or crystalline.

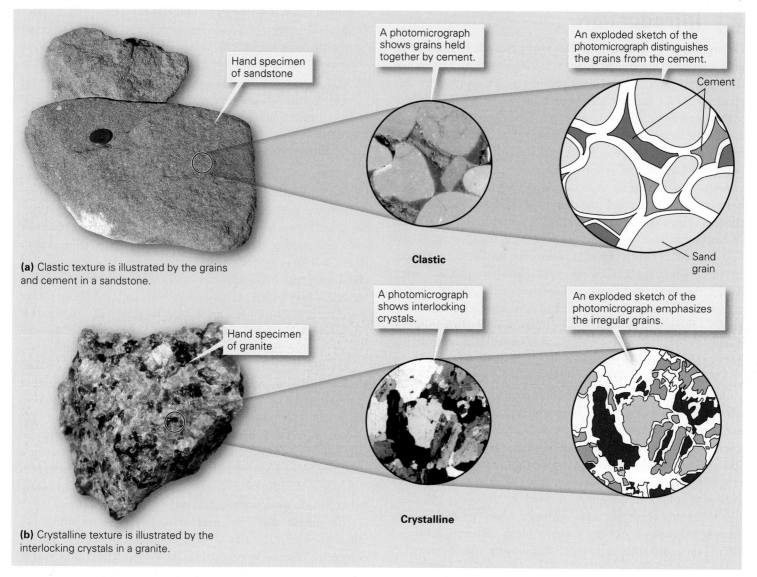

A photomicrograph shows grains held together by cement.

An exploded sketch of the photomicrograph distinguishes the grains from the cement.

Cement

Sand grain

Hand specimen of sandstone

Clastic

(a) Clastic texture is illustrated by the grains and cement in a sandstone.

A photomicrograph shows interlocking crystals.

An exploded sketch of the photomicrograph emphasizes the irregular grains.

Hand specimen of granite

Crystalline

(b) Crystalline texture is illustrated by the interlocking crystals in a granite.

approach, geologists recognize three basic groups: (1) **igneous rocks**, which form by the freezing (solidification) of molten rock (**Fig. A.3a**); (2) **sedimentary rocks**, which form either by the cementing together of fragments (grains) broken off pre-existing rocks or by the precipitation of mineral crystals out of water solutions at or near the Earth's surface (**Fig. A.3b**); and (3) **metamorphic rocks**, which form when preexisting rocks change character in response to a change in pressure and temperature conditions (**Fig. A.3c**). Metamorphic change occurs in the solid state, which means that it does not require melting. In the context of modern plate tectonics theory, different rock types form in different geologic settings, as we discuss in succeeding chapters (**Fig. A.4**).

Each of the three groups contains many different individual rock types, distinguished from one another by physical characteristics.

> *Grain size*: The dimensions of individual "grains" (here used in a general sense to mean fragments or crystals) in a rock may be measured in millimeters or centimeters. Some grains are so small that they can't be seen without a microscope, whereas others are as big as a fist or larger. Some grains are **equant**, meaning that they have the same dimensions in all directions; some are **inequant**, meaning that the dimensions are not the same in all directions (**Fig. A.5a, b**). In some rocks, all the grains are the same size, whereas other rocks contain a variety of grain sizes.

> *Composition*: A rock is a mass of chemicals. The term rock composition refers to the proportions of different chemicals making up the rock. The proportion of chemicals, in turn, affects the proportion of different minerals constituting the rock.

FIGURE A.2 Types of rock exposures.

(a) Outcrops are natural rock exposures. These outcrops rise as cliffs above the forest of the Rocky Mountains in Colorado.

(c) By blasting into the ground to produce a more level grade for roads, highway engineers produce roadcuts.

(b) In arid (dry) climates, a lack of vegetation leaves outcrops unobscured.

(d) Stream cuts form where flowing water grinds into the land and strips away soil and vegetation.

> *Texture*: This term refers to the arrangement of grains in a rock, that is, the way grains connect to one another and whether or not inequant grains are aligned parallel to each other. The concept of rock texture will become easier to grasp as we look at different examples of rocks in the following chapters.

> *Layering*: Some rock bodies appear to contain distinct layering, defined either by bands of different compositions or textures, or by the alignment of inequant grains so that they trend parallel to each other. Different types of layering occur in different kinds of rocks. For example, the layering in sedimentary rocks is called **bedding**, whereas the layering in metamorphic rocks is called **metamorphic foliation** (**Fig. A.6a, b**).

Each distinct rock type has a name. Names come from a variety of sources. Some come from the dominant component making up the rock, some from the region where the rock was first discovered or is particularly abundant, some from a root word of Latin origin, and some from a traditional name used by people in an area where the rock is found. All told, there are hundreds of different rock names, though in this book we will introduce only about 30.

A.4 Studying Rock

Outcrop Observations

The study of rocks begins by examining a rock in an outcrop. If the outcrop is big enough, such an examination will reveal relationships between the rock you're interested in and the rocks around it, and will allow you to detect layering. Geologists carefully record observations about an outcrop, then break off a **hand specimen**, a fist-sized piece, that they can examine more closely with a hand lens (magnifying glass). Observation with a hand lens enables geologists to identify sand-sized or

FIGURE A.3 Examples of the three major rock groups.

(a) Lava (molten rock) freezes to form igneous rock. Here, the molten tip of a brand-new flow still glows red. Older flows are already solid.

(b) Sand, formed from grains eroded off the rock cliffs, collects on the beach. If buried and turned to rock, it becomes layers of sandstone, such as those making up cliffs.

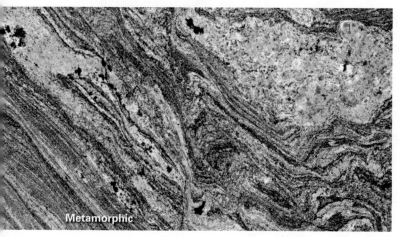

(c) Metamorphic rock forms when preexisting rocks endure changes in temperature and pressure and/or are subjected to shearing, stretching, or shortening. New minerals and textures form.

larger grains, and may enable them to describe the texture of the rock.

Thin-Section Study

Geologists often must examine rock composition and texture in minute detail in order to identify a rock and develop a hypothesis for how it formed. To do this, they take a specimen back to the lab, make a very thin slice (about 0.03 mm thick, the thickness of a human hair) and mount it on a glass slide (**Fig. A.7a–c**). They study the resulting **thin section** with a petrographic microscope (*petro* comes from the Greek word for rock). A petrographic microscope differs from an ordinary microscope in that it illuminates the thin section with transmitted polarized light. This means that the illuminating

FIGURE A.4 A cross section illustrating various geologic settings in which rocks form.

| Deep ocean | Passive Margin | Rift | | Mountain belt |

Metamorphic
rock formation

Erosion

Erosion

Sedimentary
rock formation

Sediment
deposition

Sedimentary
rock formation

Sedimentary
rock formation

light beam first passes through a special polarity filter that makes all the light waves in the beam vibrate in the same plane, and then the light passes up through the thin section and then up through another polarizing filter. An observer looks through the thin section as if it were a window. When illuminated with transmitted polarized light, and viewed through two polarizing filters, each type of mineral grain displays a unique suite of colors (**Fig. A.7d**). The specific color the observer sees depends on both the identity of the grain and its orientation with respect to the waves of polarized light.

The brilliant colors and strange shapes in a thin section viewed in polarized light rival the beauty of an abstract painting or stained glass. By examining a thin section with a petrographic microscope, geologists can identify most of the minerals constituting the rock and can describe the way in which the grains connect to each other. They can make a record of the image by using a camera. A photograph taken through a petrographic microscope is called a **photomicrograph**.

FIGURE A.5 Describing grains in rock.

(a) Grains in rock come in a variety of shapes. Some are equant, whereas some are inequant. In this example of metamorphic rock, inequant grains align to define a foliation.

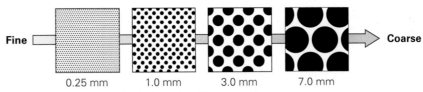

Fine | Coarse

0.25 mm   1.0 mm   3.0 mm   7.0 mm

(b) Geologists define grain size by using this comparison chart.

High-Tech Analytical Equipment

Beginning in the 1950s, high-tech electronic instruments became available that enabled geologists to examine rocks on an even finer scale than is possible with a petrographic microscope. Modern research laboratories typically boast instruments such as electron microprobes, which can focus a beam of electrons on a small part of a grain to create a signal that defines the chemical composition of the mineral (**Fig. A.8**); mass spectrometers, which analyze the proportions of atoms with different atomic weights contained in a rock; and X-ray diffractometers, which identify minerals by measuring how X-ray beams interact with crystals. Such instruments, in conjunction with optical examination, can provide geologists with highly detailed characterizations of rocks, which in turn help them understand how the rocks formed and where the rocks came from. This information enables geologists to use the study of rocks as a basis for deciphering Earth history.

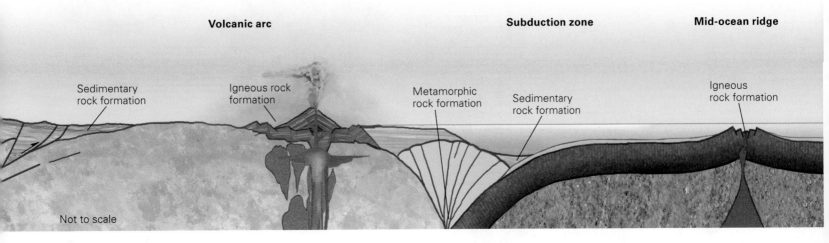

Volcanic arc   Subduction zone   Mid-ocean ridge

Sedimentary rock formation   Igneous rock formation   Metamorphic rock formation   Sedimentary rock formation   Igneous rock formation

Not to scale

FIGURE A.6 Layering in rock.

(a) Bedding in a sedimentary rock, here defined by alternating layers of coarser and finer grains, as exposed on a cliff along an Oregon beach. Older beds were tilted before younger ones were deposited.

(b) Foliation in this outcrop of metamorphic rock in Ontario, Canada is defined by alternating layers of light and dark minerals.

FIGURE A.7 Studying rocks in thin section.

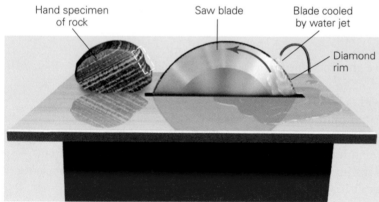

(a) Using a special saw, a geologist cuts a thin chip of a rock specimen.

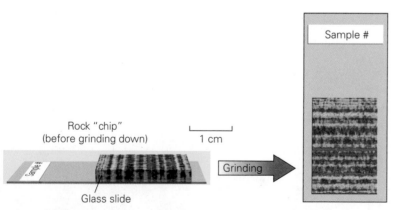

(b) The geologist glues the chip to a glass slide and grinds it down until it is so thin that light can pass through it.

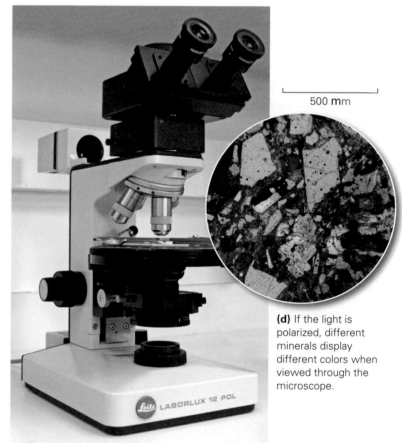

(d) If the light is polarized, different minerals display different colors when viewed through the microscope.

(c) With a petrographic microscope, it's possible to view thin sections with light that shines through the sample from below.

FIGURE A.8 An electron microprobe uses a beam of electrons to analyze the chemical composition of minerals.

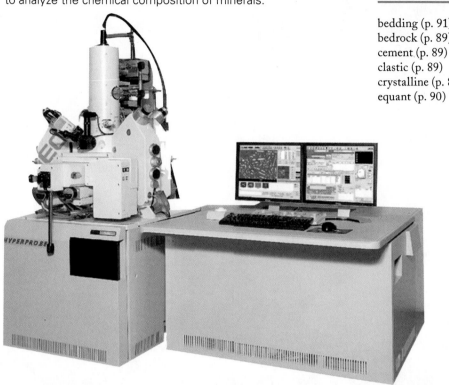

Key Terms

bedding (p. 91)
bedrock (p. 89)
cement (p. 89)
clastic (p. 89)
crystalline (p. 89)
equant (p. 90)

hand specimen (p. 91)
igneous rock (p. 90)
inequant (p. 90)
metamorphic foliation (p. 91)
metamorphic rock (p. 90)
outcrop (p. 89)
photomicrograph (p. 93)
rock (p. 89)
sedimentary rock (p. 90)
thin section (p. 92)

ANOTHER VIEW This quarry, in northwestern Italy, provides blocks of pure white marble, some of which have been carved into beautiful sculptures. It also provides a view into the bedrock that lies beneath rugged peaks.

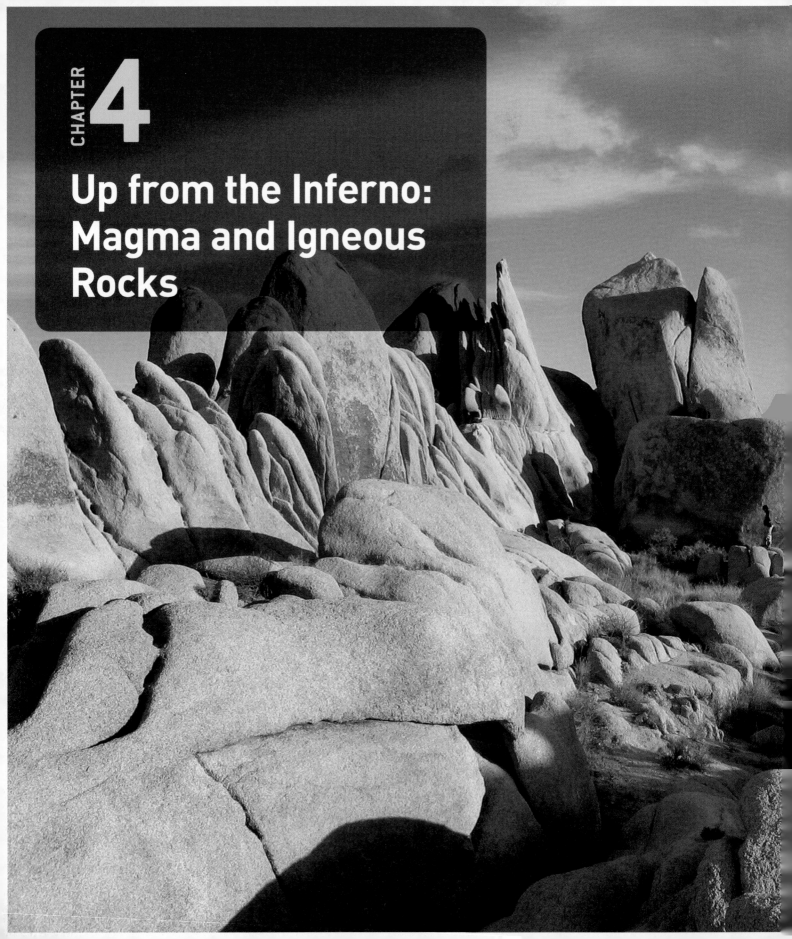

Up from the Inferno: Magma and Igneous Rocks

Massive blocks of granite crop out in Joshua Tree National Monument, California. The rock formed by the slow solidification of magma 15 km underground, about 140 million years ago.

To their grey heights they rise,
The basalt crags thus far into blue air...
Archaic columns of fire frozen to stone.

—Kathleen Raine (British poet, 1908–2003)

4.1 Introduction

Every now and then, an incandescent liquid—hot molten rock, or "melt"—fountains from a crater or crack on the big island of Hawaii. Hawaii is a **volcano**, a vent at which melt from inside the Earth spews onto the planet's surface, an event called a *volcanic eruption*. Geologists refer to melt that has emerged at the surface as **lava**. Some lava pools around the vent, while some runs down the mountainside as a syrupy red-yellow stream called a **lava flow**. Near its source, lava on Hawaii has a temperature of about 1150°C and moves swiftly, cascading over escarpments at speeds of up to 60 km per hour (**Fig. 4.1a**). At the base of the mountain, the lava flow slows but advances nonetheless, engulfing roads, houses, or vegetation in its path (**Fig. 4.1b**). As the flow cools, its surface darkens and crusts over, occasionally breaking to reveal the hot, sticky mass that continues to ooze within. Finally, the flow stops moving entirely, and within days or weeks the once red-hot melt has become a hard, black solid through and through (**Fig. 4.1c, d**). New **igneous rock**, rock made by the freezing (solidifying) of a melt, has formed. Considering the fiery heat of the melt from which igneous rocks solidify, the name igneous—from the Latin *ignis*, meaning fire—makes sense. Igneous rocks are very common on Earth. They make up all of the oceanic crust and much of the continental crust.

It may seem strange to speak of "freezing" in the context of forming rock, for most people think of freezing as the transformation of liquid water to solid ice when the temperature drops below 0°C (32°F). Nevertheless, the freezing of liquid melt to form solid igneous rock represents the same phenomenon, solidification of a liquid, except that igneous rocks freeze at high temperatures—between 650°C and 1100°C. To put such temperatures in perspective, keep in mind that home ovens attain a maximum temperature of only 260°C (500°F).

Rock that forms by the freezing of lava above ground, after it spills out (extrudes) onto the surface of the Earth and comes into contact with the atmosphere or ocean, is **extrusive igneous rock** (**Fig. 4.2a**). Extrusive igneous rock includes both solid lava flows, formed when streams or mounds of lava solidify on the surface of the Earth, and deposits of pyroclastic debris (from the Greek word *pyro*, meaning fire). Such debris includes **volcanic ash**, consisting of fine particles of glass that form when a spray of lava erupts into the air and freezes instantly. Some ash billows up several kilometers into the sky above a volcano and eventually drifts down in a snow-like ash fall. But some ash rushes down the side of the volcano in a scalding avalanche called an ash flow. Pyroclastic debris also includes larger fragments formed when clots of

FIGURE 4.1 Formation and evolution of lava flows.

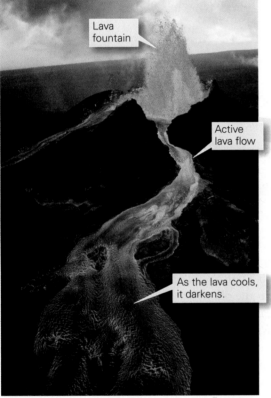

Lava fountain

Active lava flow

As the lava cools, it darkens.

(a) Lava erupts as a fountain from a volcanic vent on Hawaii. A fast-moving river of lava then flows downslope.

Smoke comes from burning vegetation.

STOP

(b) At a distance from the vent, the lava has completely crusted over with new rock, but the interior of the flow remains molten.

Time

(c) Eventually, the flow cools completely and becomes a layer of new rock. This flow engulfed a road on Hawaii.

The reddish color comes from weathering.

(d) Over time, many lava flows can accumulate one on top of another to build a large volcano. A canyon cut into the volcano exposes dozens of ancient flows.

lava freeze in the air, or when the force of the eruption blasts apart pre-existing rock of a volcano (**Fig. 4.2b**). We'll provide further detail about pyroclastic debris in Chapter 5.

The spectacle of a Hawaiian volcano erupting may give the impression that the formation of igneous rocks happens exclusively at the Earth's surface. But in fact, a vastly greater volume of igneous rock forms by solidification of molten rock underground and out of view. Geologists refer to underground melt as **magma**. Magma pushes its way, or

"intrudes," into pre-existing rock (called wall rock), where it may accumulate into an irregularly shaped mass called a **magma chamber**, rise to form a chimney-like column, or inject into cracks to form tabular underground sheets. When magma in such intrusions solidifies underground, it becomes **intrusive igneous rock**.

A great variety of igneous rocks exist on Earth. To understand why and how these rocks form, and why there are so many different kinds, we first discuss why magma forms, why it rises, how it flows, and how it freezes in intrusive and extrusive environments. We then look at the scheme that geologists use to classify igneous rocks.

FIGURE 4.2 The intrusive and extrusive realms.

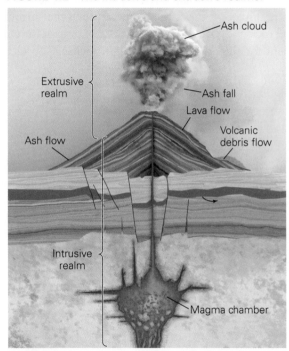

(a) The intrusive realm lies underground and the extrusive realm lies above ground. Lava flows, as well as various types of ash eruptions, all produce extrusive rocks.

(b) Broken blocks of lava were blasted out of this Alaskan volcano and now litter its slopes.

> ### Take-Home Message
> Molten rock underground is called magma, whereas molten rock that has come out of a vent at the Earth's surface is lava. Solidification of magma produces intrusive rocks. Solidification of lava, either in flows on the surface, or as fragments cooled in the air, produces extrusive igneous rock.

4.2 Why Does Magma Form, and What Is It Made of?

It's Hot Inside the Earth

Where does the heat that can cause the production of magma come from? As we discussed in Chapter 1, some of the Earth's internal heat is a relict of the planet's formation. In fact, during the first 700 million years or so of its existence, the Earth was very hot, and at times may even have been largely molten. But our planet has had a long time to cool since then, and probably would have become too cool to melt at all were it not for the presence of radioactive elements. Every time a radioactive element decays, it generates new heat. The Earth produces enough radioactive heat to keep its inside quite hot.

Causes of Melting

Even though the Earth is very hot inside, the popular image that the crust floats on a sea of molten rock is wrong. The crust and the mantle of this planet are almost entirely solid. Magma forms only in special places where pre-existing solid rock undergoes melting. Below, we describe conditions that lead to melting. We'll briefly note the settings, in the context of plate tectonics,

> **Did you ever wonder...**
> whether Earth's crust floats on a magma sea?

in which melting conditions develop, but will wait until the end of this chapter to characterize specific types of igneous rocks that form at these settings.

Melting due to a decrease in pressure (decompression).

Beneath typical oceanic crust, temperatures comparable to those of lava occur in the upper mantle (**Fig. 4.3a**). But even though the upper mantle is very hot, its rock stays solid because it is also under great pressure from the weight of overlying rock, and pressure prevents atoms from breaking free of solid mineral crystals. Because pressure prevents melting, a decrease in pressure can permit melting. Thus, if the pressure affecting hot mantle rock decreases while the temperature remains unchanged, magma forms. This kind of melting, called *decompression melting*, occurs where hot mantle rock rises to shallower depths in the Earth. Such movement occurs in mantle plumes, beneath rifts, and beneath mid-ocean ridges (**Fig. 4.3b**).

Melting as a result of the addition of volatiles.

Magma also forms at locations where chemicals called volatiles mix with hot mantle rock. Volatiles, as noted in Chapter 1, are substances such as water (H_2O) and carbon dioxide (CO_2) that evaporate easily and can exist in gaseous forms at the Earth's surface. When volatiles mix with hot, dry rock, they help break chemical bonds so that the rock begins to melt (**Fig. 4.4a**). In effect, adding volatiles decreases a rock's melting temperature. Melting due to addition of volatiles is sometimes called *flux melting*.

Melting as a result of heat transfer from rising magma.

When magma from the mantle rises up into the crust, it brings heat with it. This heat raises the temperature of the surrounding crustal rock and, in some cases, the rise in temperature may be sufficient for the crustal rock to begin melting. To picture

FIGURE 4.3 The concept of decompression melting.

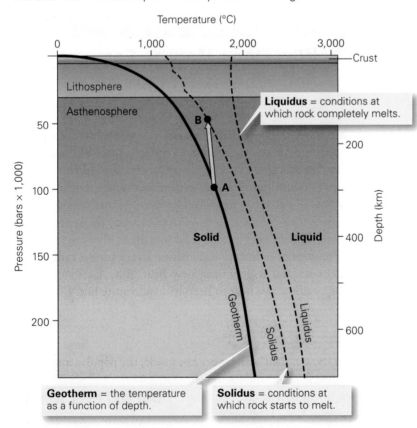

(a) Decompression melting takes place when the pressure acting on hot rock decreases. As this graph of pressure and temperature conditions in the Earth shows, when rock rises from point A to point B, the pressure decreases a lot, but the rock cools only a little, so the rock begins to melt.

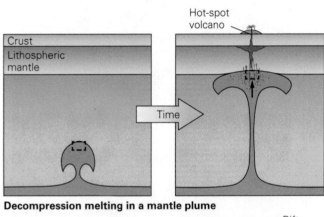

Decompression melting in a mantle plume

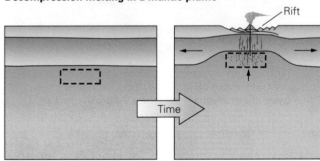

Decompression melting beneath a rift

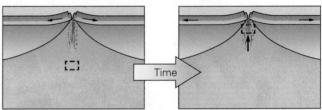

Decompression melting beneath a mid-ocean ridge

(b) The conditions leading to decompression melting occur in several different geologic environments. In each case, a volume of hot asthenosphere (outlined by dashed lines) rises to a shallower depth, and magma (red dots) forms.

the process, imagine injecting hot fudge into ice cream; the fudge transfers heat to the ice cream, raises its temperature, and causes it to melt (**Fig. 4.4b**). We call such melting *heat-transfer melting*, because it results from the transfer of heat from a hotter material to a cooler one.

The Major Types of Magma

All magmas contain silica, a compound of silicon and oxygen. But magmas also contain varying proportions of other elements such as aluminum (Al), calcium (Ca), sodium (Na), potassium (K), iron (Fe), and magnesium (Mg); each of these ions also bonds to oxygen to form a metal-oxide compound. Because magma is a liquid, its molecules do not lie in an orderly crystalline lattice but are grouped instead in clusters or short chains, relatively free to move with respect to one another.

Geologists distinguish between "dry" magmas, which contain no volatiles, and "wet" magmas, which do. In fact, wet magmas include up to 15% dissolved volatiles such as water, carbon dioxide, nitrogen (N_2), hydrogen (H_2), and sulfur dioxide (SO_2). These volatiles come out of the Earth at volcanoes in the form of gas. Usually, water constitutes about half of the gas erupting at a volcano. Thus, magma contains not only the molecules that constitute solid minerals in rocks but also the molecules that become water or air.

Magmas differ from one another in terms of the proportions of chemicals that they contain. Geologists distinguish four major compositional types depending, overall, on the proportion of silica (SiO_2) relative to other metal oxides (**Table 4.1**). **Mafic magma** contains a relatively high proportion of iron oxide (FeO or Fe_2O_3) and magnesium oxide (MgO) relative to silica. The "ma-" in the word stands for magnesium, and the

TABLE 4.1	The Four Categories of Magma
Felsic (or silicic) magma	66–76% silica*
Intermediate magma	52–66% silica
Mafic magma	45–52% silica
Ultramafic magma	38–45% silica

*The numbers provided are "weight percent," meaning the proportion of the magma's weight that consists of silica (SiO_2).

"-fic" comes from the Latin word for iron. **Ultramafic magma** has an even higher proportion of magnesium and iron oxides, relative to silica. *Felsic magmas* have a relatively high proportion of silica, relative to magnesium and iron oxides. (Occasionally, geologists use the term "silicic" interchangeably with felsic.) *Intermediate magma* gets its name because its composition is partway between mafic and felsic.

Why are there so many kinds of magma? Several factors control magma composition, including those described below.

> *Source rock composition:* The composition of a melt reflects the composition of the solid from which it was derived. Not all magmas form from the same source rock, so not all magmas have the same composition.

> *Partial melting:* Under the temperature and pressure conditions that occur in the Earth, only about 2% to 30% of an original rock can melt to produce magma at a given location—the temperature at sites of magma production simply never gets high enough to melt the entire original rock before the magma has a chance to migrate away from its source.

FIGURE 4.4 Flux melting and heat-transfer melting.

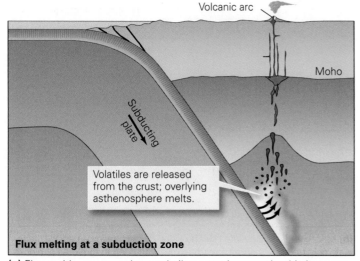

(a) Flux melting occurs where volatiles enter hot mantle; this happens at subduction zones.

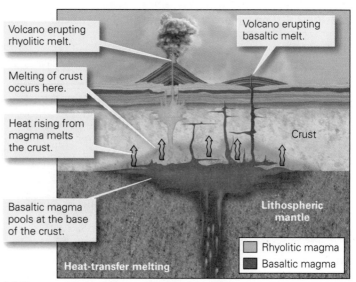

(b) Heat-transfer melting occurs when rising magma brings heat up with it and melts overlying or surrounding rock. (Not to scale.)

Partial melting refers to the process by which only part of an original rock melts to produce magma (**Fig. 4.5a**). Magmas formed by partial melting are more felsic than the original rock from which they were derived. For example, partial melting of an ultramafic rock produces a mafic magma.

> *Assimilation*: As magma sits in a magma chamber before completely solidifying, it may incorporate chemicals dissolved from the wall rocks of the chamber or from blocks that detached from the wall and sank into the magma (**Fig. 4.5b**). This process is called contamination or **assimilation**.

> *Magma mixing*: Different magmas formed in different locations from different sources may enter a magma chamber. In some cases, the originally distinct magmas mix to create a new, different magma. Thoroughly mixing a felsic magma with a mafic magma in equal proportions produces an intermediate magma.

Take-Home Message

The Earth is hot inside. Even so, the crust and mantle are solid, except in special places where pre-existing solid rock undergoes melting. Melting can be triggered by a decrease in pressure, addition of volatiles, and/or injection of hot magma from deeper below. Geologists classify magma based on its composition.

4.3 Movement and Solidification of Molten Rock

If magma stayed put once it formed, new igneous rocks would not develop in or on the crust. But it doesn't stay put; magma tends to move upward, away from where it formed. In some cases, it reaches the Earth's surface and erupts at a volcano. This movement is a key component of the Earth System, because it transfers material from deeper parts of the Earth upward and provides the raw material from which new rocks and the atmosphere and ocean form. Eventually, magma freezes and transforms into a new solid rock.

Why Does Magma Rise?

Magma rises for two reasons. First, buoyancy drives magma upward just as it drives a wooden block up through water, because magma is less dense than the surrounding rock. Second, magma rises because the weight of overlying rock creates pressure at depth that literally squeezes magma upward. The same process happens when you step into a puddle barefoot and mud squeezes up between your toes.

What Controls the Speed of Flow?

Viscosity, or resistance to flow, affects the speed with which magmas or lavas move. Magmas with low viscosity flow more easily than those with high viscosity, just as water flows more easily than molasses. Viscosity depends on temperature, volatile content, and silica content. Hotter magma is less viscous than cooler magma, just as hot tar is less viscous than cool tar, because thermal energy breaks bonds and allows atoms to move more easily. Similarly, magmas or lavas containing more volatiles are less viscous than dry (volatile-free) magmas, because the volatiles also tend to break apart silicate molecules and may also form gas bubbles. Mafic magmas are less viscous than felsic magmas, because silicon-oxygen tetrahedra tend to link together in magma to create long molecular chains that can't move past each

FIGURE 4.5 Phenomena that can affect the composition of magma.

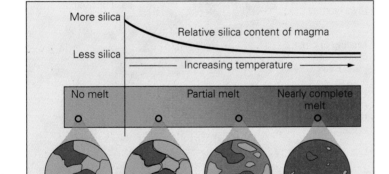

(a) Partial melting: The first-formed melt will be richer in silica than the original rock. As melting continues, magma becomes increasingly mafic.

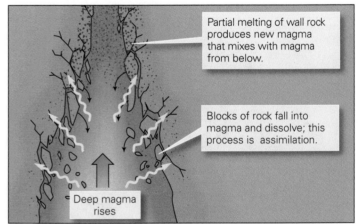

(b) Mixing and assimilation: Heat provided by deep magma partially melts wall rock; the new magma may then mix with deep magma. Also, blocks of wall rock can dissolve (assimilate) in the deep magma, and the wall rock may chemically react with the magma.

other easily, and there are more of these chains in a felsic magma than in a mafic magma. Thus, hotter mafic lavas have relatively low viscosity and flow in thin sheets over wide regions, but cooler felsic lavas are highly viscous and may clump into a dome-like mound at the volcanic vent (**Fig. 4.6a, b**).

Transforming Melt into Rock

If a melt stayed at its point of origin, and nothing in its surroundings changed, it would stay molten. But melts don't last forever. Rather, they eventually solidify or "freeze." This process happens, in some cases, because volatiles escape from the melt, so that the freezing temperature rises—if the melt's temperature stays the same but its freezing temperature rises, it will solidify. Most often, however, freezing takes place simply when melt cools below its freezing temperature. Temperature decreases upward, toward the Earth's surface, so magma enters a cooler environment automatically as it rises. If it is trapped underground as an intrusion, it slowly loses heat to surrounding wall rock, drops below its freezing temperature, and solidifies. If melt extrudes as lava at the ground surface, it cools in contact with air or water.

The time it takes for a magma to cool depends on how fast it is able to transfer heat into its surroundings. To see why, think about the process of cooling coffee. If you pour hot coffee into a thermos bottle and seal it, the coffee stays hot for hours; because it's insulated, the coffee in the thermos loses heat to the air

outside only very slowly. Like the thermos bottle, surrounding wall rock acts as an insulator in that it transports heat away from a magma only very slowly, so magma underground (in an intrusive environment) cools slowly. In contrast, if you spill coffee on a table, it cools quickly because it loses heat to the cold air. Similarly, lava that erupts at the ground surface cools quickly because the air or water surrounding it can conduct heat away quickly.

Three factors control the cooling time of magma that freezes below the surface in the intrusive realm.

> *The depth of intrusion*: Magma intruded deep in the crust, where it is surrounded by warm wall rock, cools more slowly than does magma intruded into cold wall rock near the ground surface.

> *The shape and size of a magma body*: Heat escapes from magma at an intrusion's surface, so the greater the surface area for a given volume of intrusion, the faster it cools. Thus, a body of magma roughly with the shape of a pancake cools faster than one with the shape of a melon. And since the ratio of surface area to volume increases as size decreases, a body of magma the size of a car cools faster than one the size of a ship (**Fig. 4.7a, b**).

> *The presence of circulating groundwater*: Water passing through magma absorbs and carries away heat, much like the coolant that flows around an automobile engine.

Changes in Magma during Cooling: Fractional Crystallization

Most people are familiar with the process of forming ice out of liquid water—cool the water to a temperature of 0°C and crystals of ice start to form. Keep the temperature cold enough for long enough and all the water becomes solid, composed entirely of one type of mineral—water ice. The process of freezing magma or lava is much more complex, because molten rock contains many different compounds, not just water, so during freezing of molten rock, many different minerals form. Further, not all of these minerals form at the same time (**Box 4.1**). To get a sense of this complexity, let's look at an example.

When a mafic magma starts to freeze, mafic (iron- and magnesium-rich) minerals such as olivine and pyroxene start to crystallize first. These solid crystals are denser than the remaining magma, so they start to sink (**Fig. 4.7c**). Some react chemically with the remaining magma as they sink, but some reach the floor of the magma chamber and become isolated from the magma. This process of sequential crystal formation and settling is called **fractional crystallization**—it progressively extracts iron and magnesium from the magma, so the remaining magma becomes more felsic. If a magma freezes completely before much fractional crystallization has occurred, the magma becomes mafic igneous rock. But freezing of a magma that has been left over after lots of fractional crystallization has occurred produces felsic igneous rock.

FIGURE 4.6 Viscosity affects lava behavior.

Lava flow — Lava fountain

(a) Mafic lava has relatively low viscosity. It can erupt in fountains, move long distances, and form thin lava flows.

Lava dome

(b) Felsic to intermediate lava is very viscous. When it erupts, it may form a mound-like lava dome around the volcano's vent.

4.4 How Do Extrusive and Intrusive Environments Differ?

With a background on how melts form and freeze, we can now introduce key features of the two settings—intrusive and extrusive—in which igneous rocks form.

FIGURE 4.7 Factors that affect the freezing of molten rock.

	Faster cooling	Slower cooling
Effect of size	For a given shape, a smaller volume cools faster.	
Effect of shape	For a given volume, a pancake shape cools faster.	

(a) The larger the ratio of its surface area to volume, the faster an intrusion cools. This ratio depends on both size and shape.

Extrusive Igneous Settings

Different volcanoes extrude molten rock in different ways. Some volcanoes erupt streams of low-viscosity lava that flood down the flanks of the volcano and then cover broad swaths of the countryside. When this lava freezes, it forms a relatively thin lava flow. Such flows may cool in days to months. In contrast, some volcanoes erupt viscous masses of lava that pile into rubbly domes. And still others erupt explosively, sending clouds of volcanic ash and debris skyward, and/or avalanches of ash tumbling down the sides of the volcano.

Which type of eruption occurs depends largely on a magma's composition and volatile content. Volatile-rich felsic lavas tend to erupt explosively and form thick ash and debris deposits (**Fig. 4.8a, b**). Mafic lavas tend to have low viscosity and spread in broad, thin flows (**Fig. 4.8c, d**). We discuss the products of extrusive eruptions in more detail in Chapter 5.

Intrusive Igneous Settings

Magma rises and intrudes into pre-existing rock by slowly percolating upward between grains and/or by forcing open cracks. The magma that doesn't make it to the surface freezes solid underground in contact with pre-existing rock and becomes intrusive igneous rock. As we noted, geologists commonly refer to the preexisting rock into which magma intrudes as wall rock. The boundary between wall rock and an intrusive igneous rock is called an intrusive contact.

Geologists distinguish among different types of intrusions on the basis of their shape. Tabular intrusions, or sheet intrusions, are planar and are of roughly uniform thickness. Most are in the range of centimeters to tens of meters thick, and tens of meters to tens of kilometers long. A **dike** is a tabular intrusion that cuts across pre-existing layering (bedding or

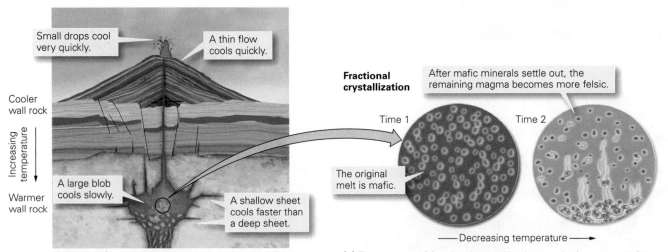

(b) The cooling rate of molten rock depends on the size and shape of the magma or lava body, and on its depth.

(c) The process of fractional crystallization results in a progressive change in magma composition during freezing.

Bowen's Reaction Series

In the 1920s, the Canadian geologist Norman L. Bowen began a series of laboratory experiments designed to determine the sequence in which silicate minerals crystallize from a melt. First, Bowen melted powdered mafic igneous rock by raising its temperature to about 1280°C. Then he cooled the melt just enough to cause part of it to solidify. Finally, he "quenched" the remaining melt by submerging it quickly in cold mercury. Quenching, which means sudden cooling to form a solid, transformed any remaining liquid into glass. The glass trapped the earlier-formed crystals within it. Bowen identified mineral crystals formed before quenching with a microscope, and he analyzed the chemical composition of the remaining glass.

After experiments at different temperatures, Bowen found that, as new crystals form, they extract certain chemicals preferentially from the melt (**Fig. Bx4.1a**). Thus, the chemical composition of the remaining melt progressively changes as the melt cools. Bowen described the specific sequence of mineral-producing reactions that take place in a cooling, initially mafic, magma. This sequence is now called **Bowen's reaction series** in his honor.

Let's examine the sequence more closely. In a cooling melt, olivine and calcium-rich plagioclase form first. This plagioclase reacts with the melt to form more, but different plagioclase; the plagioclase formed at a later stage contains more sodium (Na). Meanwhile, some olivine crystals react with the remaining melt to produce pyroxene, which may encase early olivine crystals or even replace them. However, some of the early olivine and Ca plagioclase crystals settle out of the melt, taking iron, magnesium, and calcium atoms with them. By this process, the remaining melt becomes progressively enriched in silica. As the melt continues to cool, plagioclase continues to form, with later-formed plagioclase having progressively more sodium than earlier-formed plagioclase. Pyroxene crystals react with melt to form amphibole, and then amphibole reacts with the remaining melt to form biotite. All the while, crystals continue to settle out, so the remaining melt

continues to become more felsic. At temperatures of between 650°C and 850°C, only about 10% melt remains, and this melt has a high silica content. At this stage, the final melt freezes, yielding quartz, K-feldspar (orthoclase), and muscovite.

On the basis of his observations, Bowen realized that there are two tracks to the reaction series. The "discontinuous" reaction series refers to the sequence olivine, pyroxene, amphibole, biotite, K-feldspar-muscovite-quartz in that each

step yields a different class of silicate mineral. The "continuous" reaction series refers to the progressive change from calcium-rich to Na-rich plagioclase: the steps yield different versions of the same mineral (**Fig. Bx4.1b**). It's important to note that not all minerals listed in the series appear in all igneous rock. For example, a mafic magma may completely crystallize before felsic minerals such as quartz or K-feldspar have a chance to form.

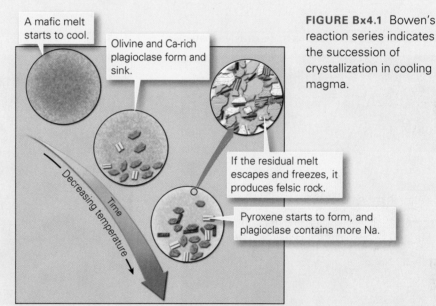

A mafic melt starts to cool.

Olivine and Ca-rich plagioclase form and sink.

If the residual melt escapes and freezes, it produces felsic rock.

Pyroxene starts to form, and plagioclase contains more Na.

Decreasing temperature

Time

FIGURE Bx4.1 Bowen's reaction series indicates the succession of crystallization in cooling magma.

(a) With decreasing temperature, fractional crystallization begins and the composition of the remaining magma becomes more felsic.

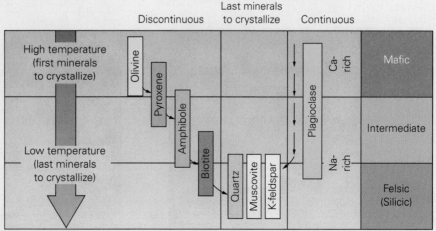

Discontinuous   Last minerals to crystallize   Continuous

High temperature (first minerals to crystallize)

Olivine

Pyroxene

Amphibole

Biotite

Quartz

Muscovite

K-feldspar

Plagioclase

Ca-rich

Na-rich

Mafic

Intermediate

Felsic (Silicic)

Low temperature (last minerals to crystallize)

(b) This chart displays the discontinuous and continuous reaction series. Rocks formed from minerals at the top of the series are mafic, whereas rocks formed from the bottom of the series are felsic.

foliation), whereas a **sill** is a tabular intrusion that injects parallel to layering (**Fig. 4.9a–d**). In places where tabular intrusions cut across rock that does not have layering, a nearly vertical, wall-like tabular intrusion is called a dike, and a nearly horizontal, tabletop-shaped tabular intrusion is called a sill. Some intrusions start to inject between layers but then dome upward, creating a blister-shaped intrusion known as a **laccolith**.

Plutons are blob-shaped intrusions that range in size from tens of meters across to tens of kilometers across (**Fig. 4.10a–e**). The intrusion of numerous plutons in a region creates a vast composite body that may be several hundred kilometers long and over 100 km wide; such immense masses of igneous rock are called **batholiths**. The rock making up the Sierra Nevada of California is a batholith formed from plutons that intruded between 145 and 80 million years ago.

Where does the space for intrusions come from? Dikes form in regions where the crust is being stretched horizontally, such as in a rift. Thus, as the magma that makes a dike forces its way up into a crack, the crust opens up sideways (**Fig. 4.11a**). Intrusion of sills occurs near the surface of the Earth, so the pressure of the magma effectively pushes up the rock above the sill, leading to uplift of the Earth's surface (**Fig. 4.11b**).

How does the space for a pluton develop? Some geologists propose that a pluton is a frozen "diapir," meaning a light-bulb-shaped blob of magma that pierced overlying rock and pushed it aside as it rose (**Fig. 4.11c**). Another explanation involves **stoping**, a process during which magma assimilates wall rock, and blocks of wall rock break off and sink into the magma (**Fig. 4.11d**). If a stoped block does not

FIGURE 4.8 Examples of eruptions and extrusive materials.

(a) This volcanic explosion produced two styles of ash eruption.

(b) Thick layers of tuff deposited by explosive eruptions in New Mexico, about 1.14 Ma. Note the highway, for scale.

(c) This volcano is producing lava flows and fountains.

(d) A stack of over 50 thin lava flows, capped by pyroclastic debris, visible from inside Mt. Vesuvius, Italy.

FIGURE 4.9 Igneous sills and dikes, examples of tabular intrusions.

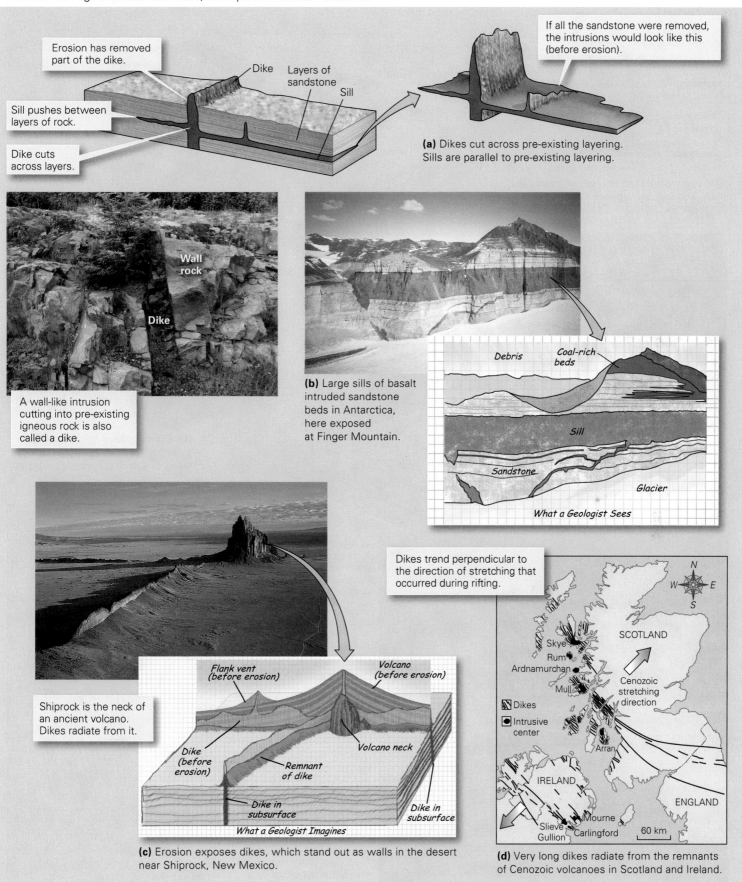

Erosion has removed part of the dike.

Sill pushes between layers of rock.

Dike cuts across layers.

Dike Layers of sandstone Sill

If all the sandstone were removed, the intrusions would look like this (before erosion).

(a) Dikes cut across pre-existing layering. Sills are parallel to pre-existing layering.

Wall rock

Dike

A wall-like intrusion cutting into pre-existing igneous rock is also called a dike.

(b) Large sills of basalt intruded sandstone beds in Antarctica, here exposed at Finger Mountain.

Debris Coal-rich beds

Sill

Sandstone Glacier

What a Geologist Sees

Dikes trend perpendicular to the direction of stretching that occurred during rifting.

Shiprock is the neck of an ancient volcano. Dikes radiate from it.

Flank vent (before erosion) Volcano (before erosion)

Dike (before erosion) Remnant of dike Volcano neck

Dike in subsurface Dike in subsurface

What a Geologist Imagines

(c) Erosion exposes dikes, which stand out as walls in the desert near Shiprock, New Mexico.

SCOTLAND

Skye Rum Ardnamurchan Mull Arran

Cenozoic stretching direction

Dikes
Intrusive center

IRELAND

ENGLAND

Slieve Gullion Mourne Carlingford 60 km

(d) Very long dikes radiate from the remnants of Cenozoic volcanoes in Scotland and Ireland.

FIGURE 4.10 Igneous plutons, "blob-shaped" intrusions.

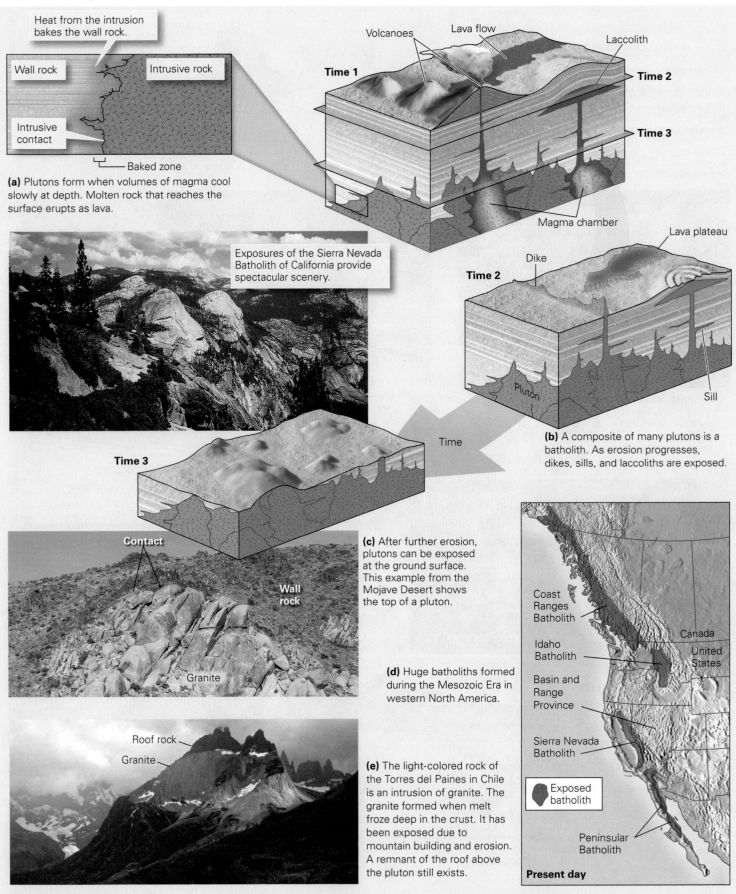

Heat from the intrusion bakes the wall rock.

Wall rock

Intrusive rock

Intrusive contact

Baked zone

(a) Plutons form when volumes of magma cool slowly at depth. Molten rock that reaches the surface erupts as lava.

Volcanoes   Lava flow   Laccolith

Time 1   **Time 2**

Time 3

Magma chamber

Exposures of the Sierra Nevada Batholith of California provide spectacular scenery.

Lava plateau

Dike

Time 2

Pluton   Sill

(b) A composite of many plutons is a batholith. As erosion progresses, dikes, sills, and laccoliths are exposed.

Time 3

Time

Contact

Wall rock

Granite

(c) After further erosion, plutons can be exposed at the ground surface. This example from the Mojave Desert shows the top of a pluton.

Coast Ranges Batholith

Canada

Idaho Batholith

United States

Basin and Range Province

Sierra Nevada Batholith

(d) Huge batholiths formed during the Mesozoic Era in western North America.

Exposed batholith

Roof rock

Granite

(e) The light-colored rock of the Torres del Paines in Chile is an intrusion of granite. The granite formed when melt froze deep in the crust. It has been exposed due to mountain building and erosion. A remnant of the roof above the pluton still exists.

Peninsular Batholith

Present day

FIGURE 4.11 Making room for an igneous intrusion.

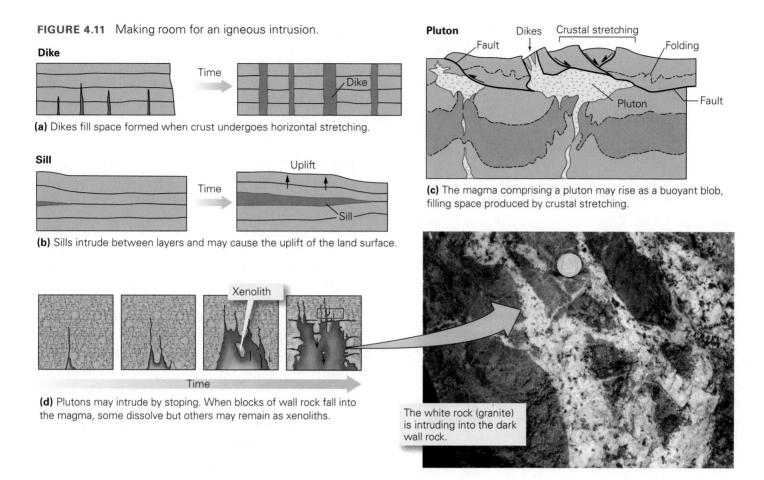

Dike

(a) Dikes fill space formed when crust undergoes horizontal stretching.

Sill

(b) Sills intrude between layers and may cause the uplift of the land surface.

(d) Plutons may intrude by stoping. When blocks of wall rock fall into the magma, some dissolve but others may remain as xenoliths.

Xenolith

Pluton

(c) The magma comprising a pluton may rise as a buoyant blob, filling space produced by crustal stretching.

The white rock (granite) is intruding into the dark wall rock.

melt entirely, but rather becomes surrounded by new igneous rock, it is a **xenolith**, after the Greek word *xeno*, meaning foreign. More recently, geologists have proposed that plutons form by injection of numerous superimposed dikes or sills, which coalesce and recrystallize to become a single, massive body.

Take-Home Message

Molten rock can extrude either as a lava flow or as pyroclastic debris. Intrusions underground juxtapose igneous rock with wall rock. Tabular intrusions (sills and dikes) are wall-like intrusions. Blob-shaped intrusions are plutons. Huge batholiths consist of many plutons.

4.5 How Do You Describe an Igneous Rock?

Characterizing Color and Texture

If you wander around a city admiring building facades, you'll find that many facades consist of igneous rock, for such rocks tend to be very durable. If you had to describe one of these rocks to a friend, what words might you use? You would

probably start by noting the rock's color. Overall, is the rock dark or light? More specifically, is it gray, pink, white, or black? Describing color may not be easy, because some igneous rocks contain many visible mineral grains, each with a different color; but even so, you'll probably be able to characterize the overall hue of the rock. Generally, the color reflects the rock's composition, but it isn't always so simple, because color may also be influenced by grain size and by the presence of trace amounts of impurities. (For example, the presence of a small amount of iron oxide gives rock a reddish tint.) Next, you would probably characterize the rock's texture. A description of igneous texture indicates whether the rock consists of glass, crystals, or fragments. If the rock consists of crystals or fragments, a description of texture also specifies the grain size. Here are the common terms for defining texture:

> *Crystalline texture*: Rocks that consist of minerals that grow when a melt solidifies interlock like pieces of a jigsaw puzzle (**Fig. 4.12a**). Rocks with such a texture are called **crystalline igneous rocks**. The interlocking of crystals in these rocks occurs because once some grains have developed, they interfere with the growth of later-formed grains. The last grains to form end up filling irregular spaces between already existing grains. Geologists distinguish subcategories of crystalline igneous rocks according to the size of the crystals. Coarse-grained (phaneritic) rocks have crystals large enough to be identified

with the naked eye. Fine-grained (aphanitic) rocks have crystals too small to be identified with the naked eye. Porphyritic rocks have larger crystals surrounded by a mass of fine crystals. In a porphyritic rock, the larger crystals are called phenocrysts, while the mass of finer crystals is called groundmass.

> *Fragmental texture*: Rocks consisting of igneous chunks and/or shards that are packed together, welded together, or cemented together after having solidified are **fragmental igneous rocks** (**Fig. 4.12a**).

> *Glassy texture*: Rocks made of a solid mass of glass, or of tiny crystals surrounded by glass, are **glassy igneous rocks** (Fig. 4.12a). Glassy rocks fracture conchoidally (**Fig. 4.12b**).

What factors control the texture of igneous rocks? In the case of nonfragmental rocks, *texture largely reflects cooling rate*. The presence of glass indicates that cooling happened so quickly that the atoms within a lava didn't have time to arrange into crystal lattices. Crystalline rocks form when a melt cools more slowly. In crystalline rocks, grain size depends on cooling time. A melt that cools rapidly, but not rapidly enough to make glass, forms fine-grained rock, because many crystals form but none has time to grow large (**Fig. 4.12c**). A melt that cools very slowly forms a coarse-grained rock, because a few crystals have time to grow large.

Because of the relationship between cooling rate and texture, lava flows, dikes, and sills tend to be composed of fine-grained igneous rock. In contrast, plutons tend to be composed of coarse-grained rock. Plutons that intrude into hot wall rock at great depth cool very slowly and thus tend to have larger crystals than plutons that intrude into cool country rock at shallow depth, where they cool relatively rapidly. Porphyritic rocks form when a melt cools in two stages. First, the melt cools slowly at depth, so that phenocrysts form. Then, the melt erupts and the remainder cools quickly, so that groundmass crystallizes around the phenocrysts.

There is, however, an exception to the standard cooling rate and grain size relationship. A very coarse-grained igneous rock called **pegmatite** doesn't necessarily cool slowly. Pegmatite contains crystals up to tens of centimeters across and occurs in dikes. Because pegmatite occurs in dikes, which generally cool quickly, the coarseness of the rock may seem surprising. Researchers have shown that pegmatites are coarse because they form from water-rich melts in which atoms can move around so rapidly that large crystals can grow very quickly.

Classifying Igneous Rocks

Because melts can have a variety of compositions and can freeze to form igneous rocks in many different environments above and below the surface of the Earth, we observe a wide spectrum of igneous rock types. We classify these according to their texture and composition. Studying a rock's texture tells us about the rate at which it cooled, as we've seen, and therefore the environment in which it formed (see **Geology at a Glance**, p. 112). Studying its composition tells us about the original source of the magma and the way in which the magma evolved before finally solidifying. Below, we introduce some of the more important igneous rock types.

Crystalline igneous rocks. The scheme for classifying the principal types of crystalline igneous rocks is quite simple. The different compositional classes are distinguished on the basis of silica content—ultramafic, mafic, intermediate, or felsic—whereas the different textural classes are distinguished according to whether the grains are coarse or fine. The chart in **Figure 4.13** gives the texture and composition of the most commonly used crystalline igneous rock names. As a rough guide, the color of an igneous rock reflects its composition: mafic rocks tend to be black or dark gray, intermediate rocks tend to be lighter gray or greenish gray, and felsic rocks tend to be light tan to pink or maroon. Figure 4.12 provides images of some of these rocks.

Note that, according to Figure 4.13, rhyolite and granite have the same chemical composition but differ in grain size. Which of these two rocks develops from a melt of felsic composition depends on the cooling rate. A felsic lava that solidifies quickly at the Earth's surface or in a thin dike or sill turns into fine-grained rhyolite; but the same magma, if solidifying slowly at depth in a pluton, turns into coarse-grained granite. A similar situation holds for mafic lavas—a mafic lava that cools quickly in a lava flow forms basalt, but a mafic magma that cools slowly forms gabbro.

Glassy igneous rocks. Glassy texture develops more commonly in felsic igneous rocks because the high concentration of silica inhibits the easy growth of crystals. But basaltic and intermediate lavas can form glass if they cool rapidly enough. In some cases, a rapidly cooling lava freezes while it still contains a high concentration of gas bubbles—these bubbles remain as open holes known as **vesicles**. Geologists distinguish among several different kinds of glassy rocks.

> **Obsidian** is a mass of solid, felsic glass. It tends to be black or brown (Fig. 4.12b). Because it breaks conchoidally, sharp-edged pieces split off its surface when you hit a sample with a hammer. Pre-industrial people worldwide used such pieces for arrowheads, scrapers, and knife blades.

Did you ever wonder...
how black glass once used for arrowheads formed?

> **Pumice** is a felsic volcanic rock that contains abundant vesicles, giving it the appearance of a sponge. Pumice forms by the quick cooling of frothy lava that resembles the head of foam in a glass of beer. In some cases, pumice contains so many air-filled pores that it can actually float on water, like styrofoam (**Fig. 4.14**).

FIGURE 4.12 Textures and types of igneous rocks.

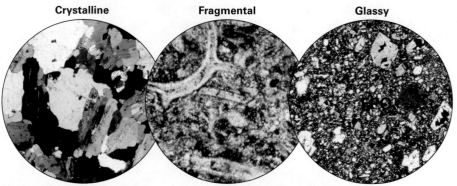

Crystalline   **Fragmental**   **Glassy**

(a) Photomicrographs of thin sections reveal the different textures of igneous rocks.

Obsidian

(b) Obsidian fractures conchoidally.

Fine grained

Coarse grained

Felsic

Increasing silica content

Mafic

Rhyolite

Andesite

Basalt

Granite, cut by pegmatite

Diorite

Gabbro

(c) Examples of igneous rocks, arranged by grain size and composition. Different environments yield different rock types.

Formation of Igneous Rocks

Igneous rock forms by the cooling of magma underground, or of lava at the surface. Igneous rocks that solidify underground are intrusive, whereas those that solidify at the surface are extrusive. The type of igneous rock that forms depends on the composition of the melt and the environment of cooling.

In the extrusive environment, melt may cool quickly and have a glassy texture. Melt that explodes into the air forms ash and other debris with fragmental texture.

Stratified volcanic tuff

Increasing silica content

Fast cooling

Slow cooling

MAFIC	FELSIC
Scoria (glassy)	Obsidian (glassy)
Basalt (fine grained)	Rhyolite (fine grained)
Gabbro (coarse grained)	Granite (coarse grained)

In the intrusive environment, crystals grow together to form an interlocking texture. Slower cooling makes coarser grains.

Cooler

Minerals in an igneous rock crystallize in succession as the melt cools.

Hotter

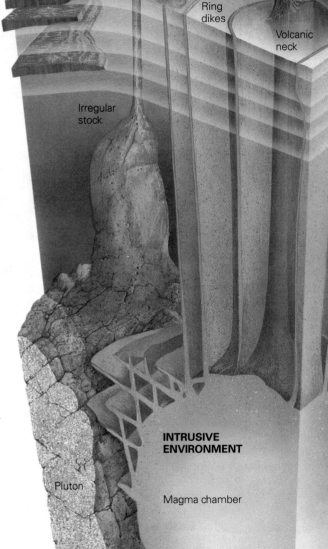

EXTRUSIVE ENVIRONMENT

Pyroclastic f

Lava flow

Dike swarm

Sills

Lava dome

Lacc

Ring dikes

Volcanic neck

Irregular stock

INTRUSIVE ENVIRONMENT

Pluton

Magma chamber

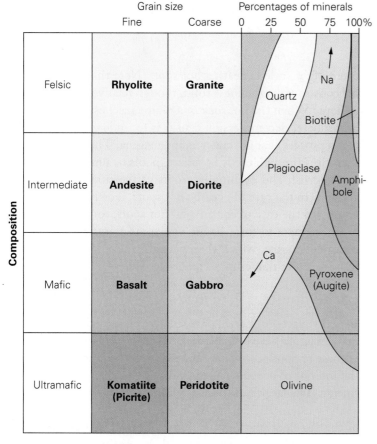

Composition	Grain size		Percentages of minerals
	Fine	Coarse	0 25 50 75 100%
Felsic	**Rhyolite**	**Granite**	Quartz / Na / Biotite
Intermediate	**Andesite**	**Diorite**	Plagioclase / Amphibole
Mafic	**Basalt**	**Gabbro**	Ca / Pyroxene (Augite)
Ultramafic	**Komatiite (Picrite)**	**Peridotite**	Olivine

FIGURE 4.13 Igneous rocks are classified based on composition and texture.

FIGURE 4.14 Pumice, a vesicle-filled volcanic rock, is so light that paper can hold it up. The vesicles it contains tend to be small.

> **Scoria** is a mafic volcanic rock that contains abundant vesicles (more than about 30%). Generally, the bubbles in scoria are bigger than those in pumice, and the rock, overall, looks darker.

Pyroclastic igneous rocks. As we have noted, when volcanoes erupt explosively, they spew out fragments of lava. Geologists refer to all such fragments as pyroclasts. Accumulations of fragmental volcanic debris are called pyroclastic deposits, and when the material in these deposits consolidates into a solid mass, due either to welding together of still-hot clasts or to cementation by minerals precipitating from water passing through, it becomes a **pyroclastic rock**. Geologists distinguish among several types of pyroclastic rocks based on grain size. Let's consider two examples.

> **Tuff** is a fine-grained pyroclastic igneous rock composed of volcanic ash. It may contain fragments of pumice.

> **Volcanic breccia** consists of larger fragments of volcanic debris that either fall through the air and accumulate, or form when a lava flow breaks into pieces.

Take-Home Message

Igneous rocks come in a variety of colors; mafic rocks tend to be darker than felsic rocks. Textures of igneous rock vary from glassy, to fine-grained, to coarse-grained. Geologists classify and assign names to igneous rocks based on texture and composition.

4.6 Plate-Tectonic Context of Igneous Activity

Earlier in this chapter, we pointed out that melting occurs only in special locations where conditions lead to decompression, addition of volatiles, and/or heat transfer. The conditions that lead to melting and, therefore, to igneous activity, can develop in four geologic settings (**Fig. 4.15**): (1) along volcanic arcs bordering oceanic trenches; (2) at hot spots; (3) within continental rifts; (4) along mid-ocean ridges. Let's look more carefully at melting and igneous rock production at these

FIGURE 4.15 The tectonic setting of igneous rocks.

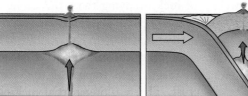
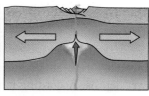

| Mantle plume and a hot-spot volcano | Subduction yields a volcanic arc. | Melting occurs beneath a mid-ocean ridge. | Melting occurs beneath a continental rift. |

settings, in the context of plate-tectonics theory, with a focus on the types of igneous rocks that may form in each setting. In Chapter 5, we'll add to the story by discussing the types of eruptions associated with different settings.

Products of Subduction

A chain of volcanoes, called a **volcanic arc** (or just an arc), forms on the overriding plate, adjacent to the deep-ocean trenches that mark convergent plate boundaries (see Chapter 2). The word "arc" emphasizes that many of these chains define a curve on a map. *Continental arcs*, such as the Andean arc of South America and the Cascade arc in the northwestern United States, grow along the edge of a continent, where oceanic lithosphere subducts beneath continental lithosphere. *Island arcs*, such as the Aleutian arc of Alaska and the Mariana arc of the western Pacific, protrude from the ocean at localities where one oceanic plate subducts beneath another. Beneath volcanic arcs, a variety of intrusions—plutons, dikes, and sills—develop, to be exposed only later, when erosion has removed the volcanic overburden. In some localities, arc-related igneous activity produces huge batholiths.

How does subduction trigger melting? Some minerals in oceanic crust rocks contain volatile compounds (mostly water). At shallow depths, volatiles are chemically bonded to the minerals. But when subduction carries crust down into the hot asthenosphere, "wet" crustal rocks warm up. At a depth of about 150 km, crust becomes so hot that volatiles separate from crustal minerals and diffuse up into the overlying asthenosphere. Addition of volatiles causes the hot ultramafic rock in the asthenosphere to undergo partial melting, a process that yields mafic magma. This magma either rises directly, to erupt as basaltic lava, or undergoes fractional crystallization before erupting and evolves into intermediate or felsic lava.

In continental volcanic arcs, not all the mantle-derived basaltic magma rises directly to the surface; some gets trapped at the base of the continental crust, and some in magma chambers deep in the crust. When this happens, heat transfers into the continental crust and causes partial melting of this crust. Because much of the continental crust is mafic to intermediate in composition to start with, the resulting magmas are intermediate to felsic in composition. This magma rises, leaving the basalt behind, and either cools higher in the crust to form plutons or rises to the surface and erupts. For this reason, granitic plutons and andesite lavas form at continental arcs.

Products of Hot Spots

As we learned in Chapter 2, most researchers think that **hot-spot volcanoes** form above plumes of hot mantle rock from deep in the mantle, though some studies suggest that some hot spots may originate due to other processes happening at shallower depths. According to the plume hypothesis, a column, or "plume" of very hot rock rises like soft plastic up through the overlying mantle beneath a hot spot. (Note that a plume does *not* consist of magma; it is solid, though relatively soft and able to flow.) When the hot rock of a plume reaches the base of the lithosphere, decompression causes it to undergo partial melting, a process that generates mafic magma. The mafic magma then rises through the lithosphere, pools in a magma chamber in the crust, and eventually erupts at the surface, forming a volcano. In the case of oceanic hot spots, mostly mafic magma erupts. In the case of continental hot spots, some of the mafic magma erupts to form basalt; but some transfers heat to the continental crust, which then partially melts itself, producing felsic magmas that erupt to form rhyolite.

Large Igneous Provinces (LIPs)

In many places on Earth, particularly voluminous quantities of mafic magma have erupted and/or intruded (**Fig. 4.16**). Some of these regions occur along the margins of continents, some in the interior of oceanic plates, and some in the interior of continents. The largest of these, the Ontong Java Oceanic Plateau of the western Pacific, covers an area of about 5,000,000 km^2 of the sea floor and has a volume of about 50,000,000 km^3. Such provinces also occur on land. It's no surprise that these huge volumes of igneous rock are called **large igneous provinces (LIPs)**. More recently, this term LIP has been applied to huge eruptions of felsic ash too.

Mafic LIPs may form when the bulbous head of a mantle plume first reaches the base of the lithosphere. More partial melting can occur in a plume head than in normal asthenosphere, because temperatures are higher in a plume head. Thus, an unusually large quantity of unusually hot basaltic magma forms in the plume head; when the magma reaches the surface, huge quantities of basaltic lava spew out of the ground. If the plume head lies beneath a rift, added decompression can lead to even more melting (**Fig. 4.17a**). The particularly hot basaltic lava that erupts at such localities has such low viscosity that it can flow tens to hundreds of kilometers across the landscape. Geoscientists refer to such flows as **flood basalts**. Flood basalts make up the bedrock of the Columbia River Plateau in Oregon and Washington (**Fig. 4.17b, c**), the Paraná Plateau in southeastern Brazil, the Karoo region of southern Africa, and the Deccan region of southwestern India.

Igneous Rocks at Rifts

Successful rifting splits a continent in two and gives birth to a new mid-ocean ridge. As the continental lithosphere thins during rifting, the weight of rock overlying the asthenosphere decreases, so pressure in the asthenosphere decreases and decompression melting produces basaltic magma, which rises into the crust. Some of this magma makes it to the surface and erupts as basalt. However, some of the magma gets trapped in the crust and transfers heat to the crust. The resulting partial melting of the crust yields felsic (silicic) magmas that erupt as rhyolite.

FIGURE 4.16 A map showing the distribution of large igneous provinces (LIPs) on Earth. The red areas are or once were underlain by immense volumes of basalt; not all of this basalt is exposed.

Thus, a sequence of volcanic rocks in a rift generally includes basaltic flows and sheets of rhyolitic lava or ash. Locally, the felsic and mafic magmas mix to form intermediate magma.

Forming Igneous Rocks at Mid-Ocean Ridges

Most igneous rocks at the Earth's surface form at mid-ocean ridges, that is, along divergent plate boundaries. Think about it—the entire oceanic crust, a 7- to 10-km-thick layer of basalt and gabbro that covers 70% of the Earth's surface, forms at mid-ocean ridges. And this entire volume gets subducted and replaced by new crust, over a period of about 200 million years.

Igneous magmas form at mid-ocean ridges for much the same reason they do at hot spots and rifts. As sea-floor spreading occurs and oceanic lithosphere plates drift away from the ridge, hot asthenosphere rises to keep the resulting space filled. As this asthenosphere rises, it undergoes decompression, which leads to partial melting and the generation of basaltic magma. As noted in Chapter 2, this magma rises into the crust and pools in a shallow magma chamber. Some cools slowly along the margins of the magma chamber to form massive gabbro, while some intrudes upward to fill vertical cracks that appear as newly formed crust splits apart (see Fig. 2.17c). Magma that cools in the cracks forms basalt dikes, and magma that makes it to the sea floor and extrudes as lava forms **pillow basalt** flows.

FIGURE 4.17 Flood basalts form when vast quantities of low-viscosity mafic lava "floods" over the landscape and freezes into a thin sheet. Accumulation of successive flows builds a flat-topped plateau.

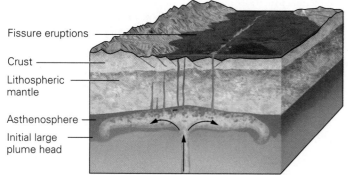

Fissure eruptions

Crust

Lithospheric mantle

Asthenosphere

Initial large plume head

(a) The plume model for forming flood basalts.

(b) Flood basalts form the layers exposed in Palouse Canyon, Washington.

Take-Home Message

The formation of igneous rocks can be understood in the context of plate-tectonics theory. Flux melting due to release of fluids from subducting slabs produces melts at convergent margins. Melting at hot spots is probably due to decompression at the top of a mantle plume. Decompression of rising asthenosphere also triggers melting beneath rifts and mid-ocean ridges. Injection of hot mantle-derived magma into the crust at rifts and convergent margins causes melting of the crust.

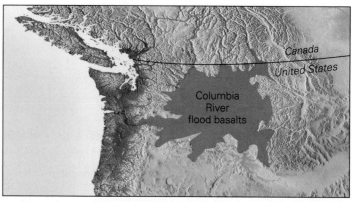

Canada

United States

Columbia River flood basalts

(c) Flood basalts underlie the Columbia River Plateau in Washington and Oregon, the dark area on this map.

CHAPTER 4 REVIEW

Chapter Summary

> Magma is liquid rock (melt) under the Earth's surface. Lava is melt that has erupted from a volcano at the Earth's surface.

> Magma forms when hot rock in the Earth partially melts. This process occurs only under certain circumstances—when the pressure decreases, when volatiles are added to hot rock, and when heat is transferred into the crust by magma rising from the mantle into the crust.

> Magma occurs in a range of compositions: felsic (silicic), intermediate, mafic, and ultramafic. The composition of magma reflects the original composition of the rock from which the magma formed and the way the magma evolves.

> Magma rises from depth because of its buoyancy and because of pressure caused by the weight of overlying rock.

> Magma viscosity depends on composition. Felsic magma is more viscous than mafic magma.

> The rate at which intrusive magma cools depends on the depth at which it intrudes, the size and shape of the magma body, and whether circulating groundwater is present. The cooling time influences the texture of an igneous rock.

> Extrusive igneous rocks form from lava that erupts out of a volcano. Intrusive igneous rocks develop from magma that freezes inside the Earth.

> Lava may solidify to form flows, or it may explode into the air to form ash.

> Intrusive igneous rocks form when magma intrudes into pre-existing rock below Earth's surface. Blob-shaped intrusions are called plutons. Tabular intrusions that cut across layering are dikes, and those that form parallel to layering are sills. Huge intrusions, made up of many plutons, are known as batholiths.

> Igneous rocks are classified according to texture and composition.

> The origin of igneous rocks can readily be understood in the context of plate tectonics. Magma forms at continental or island volcanic arcs along convergent margins, mostly because of the addition of volatiles to the asthenosphere above the subducting slab. Igneous rocks form at hot spots, owing to the decompression melting of a rising mantle plume. Igneous rocks form at rifts as a result of decompression melting of the asthenosphere below the thinning lithosphere or heat transfer from mantle melts into crustal rocks. Igneous rocks form along mid-ocean ridges because of decompression melting of the rising asthenosphere.

Key Terms

assimilation (p. 102)	glassy igneous rock (p. 110)	magma (p. 99)	solidus (p. 100)
batholith (p. 106)	hot-spot volcano (p. 114)	magma chamber (p. 99)	stoping (p. 106)
Bowen's reaction series (p. 105)	igneous rock (p. 97)	obsidian (p. 110)	tuff (p. 113)
crystalline igneous rock (p. 109)	intrusive igneous rock (p. 99)	partial melting (p. 102)	ultramafic magma (p. 101)
dike (p. 104)	laccolith (p. 106)	pegmatite (p. 110)	vesicle (p. 110)
extrusive igneous rock (p. 98)	large igneous province (LIP) (p. 114)	pillow basalt (p. 115)	viscosity (p. 102)
flood basalt (p. 114)	lava (p. 97)	pluton (p. 106)	volcanic arc (p. 114)
fractional crystallization (p. 103)	lava flow (p. 97)	pumice (p. 110)	volcanic ash (p. 98)
fragmental igneous rock (p. 110)	liquidus (p. 100)	pyroclastic rock (p. 113)	volcanic breccia (p. 113)
geotherm (p. 100)	mafic magma (p. 101)	scoria (p. 113)	volcano (p. 97)
		sill (p. 106)	xenolith (p. 109)

Review Questions

1. How is the process of freezing magma similar to that of freezing water? How is it different?

2. What is the source of heat in the Earth? How did the first igneous rocks on the planet form?

3. Describe the three processes that are responsible for the formation of magmas.

4. Why are there so many different types of magmas? Does partial melting produce magma with the same composition as the parent rock from which it was derived?

5. Why do magmas rise from depth to the surface of the Earth?

6. What factors control the viscosity of a melt?

Every chapter of SmartWork contains active learning exercises to assist you with reading comprehension and concept mastery. This chapter also features:

> Interactive exercises on lava composition.

> A video exercise addressing lava composition.

> A What a Geologist Sees exercise on magma viscosity and mineral formation in igneous rocks.

---

7. What factors control the cooling time of a magma within the crust?

8. What is the difference between a sill and a dike and how do both differ from a pluton?

9. How does grain size reflect the cooling time of a magma?

10. What does the mixture of grain sizes in a porphyritic igneous rock indicate about its cooling history?

11. Describe the way magmas are produced in subduction zones.

12. What process in the mantle may be responsible for causing hot-spot volcanoes to form?

13. Describe how magmas are produced at continental rifts. Why can you find both basalt and rhyolite in such settings?

14. What is a large igneous province (LIP)? How might the formation of LIPs have affected the Earth System?

15. Why does melting take place beneath the axis of a mid-ocean ridge?

On Further Thought

16. If you look at the Moon, even without a telescope, you see broad areas where its surface appears relatively darker and smoother. These areas are individually called *mare* (plural: *maria*), from the Latin word for sea. The term is misleading, for they are not bodies of water but rather plains of igneous rock formed after huge meteors struck the Moon and formed very deep craters. These impacts occurred early in the history of the Moon, when its interior was warmer. With this background information in mind, propose a cause for the igneous activity, and suggest the type of igneous rock that fills the mare. (*Hint*: Think about how the presence of a deep crater affects pressure in the region below the crater, and think about the viscosity of a magma that could spread over such a broad area.)

17. The Cascade volcanic chain of the northwestern United States is only about 800 km long (from the southernmost volcano in California to the northernmost one in Washington State). The volcanic chain of the Andes is several thousand kilometers long. Look at a map showing the Earth's plate boundaries, and explain why the Andes volcanic chain is so much longer than the Cascade volcanic chain.

SEE FOR YOURSELF D... ## Igneous Rocks

Download *Google Earth*™ from the Web in order to visit the locations described below (instructions appear in the Preface of this book). You'll find further locations and associated active-learning exercises on Worksheet D of our **Geotours Workbook**.

Granite of the Sierra Nevada Batholith
Latitude   37°45′18.01″N,
Longitude   119°32′21.00″W
(oblique, looking east)

This view looks at the valley of Yosemite National Park, with the famous climbing peak Half Dome on the right. The bedrock in this view consists of granite, part of a batholith that intruded during the Mesozoic either beneath an island or along a convergent margin.

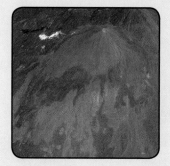

Izalco Volcano, El Salvador
Latitude   13°48′50.40″N,
Longitude   89°37′57.74″W
(oblique, looking east)

This volcano was active almost continuously from 1770 to 1958. Several basalt lava flows spread down the slopes into the green jungle. Younger flows have a darker color. The gray covering near the summit consists of pyroclastic debris.

The Wrath of Vulcan: Volcanic Eruptions

Chapter Objectives

By the end of this chapter you should know . . .

> the nature of the great variety of materials (such as lava, pyroclastic debris, and gas) that erupt at volcanoes.

> how volcanic eruptions can differ from one another, so that some yield rivers of lava, whereas others produce catastrophic explosions.

> the various hazards to life and environment that can result from volcanic eruptions.

> that, in some cases, impending eruptions can be predicted, allowing people to take precautions.

> how volcanoes may affect climate, evolution, and perhaps the future of civilizations.

Glowing waves rise and flow, burning all life on their way, and freeze into black, crusty rock which adds to the height of the mountain and builds the land, thereby adding another day to the geologic past. . . . I became a geologist forever, by seeing with my own eyes: the Earth is alive!

—Hans Cloos (1886–1951),
on seeing the eruption of Mt. Vesuvius (Italy)

5.1 Introduction

Every few hundred years, one of the hills on Vulcano, an island in the Mediterranean Sea off the western coast of Italy, rumbles and spews out molten rock, glassy cinders, and dense "smoke" (actually a mixture of various gases, fine ash, and very tiny liquid droplets). Ancient Romans thought that such eruptions happened when Vulcan, the god of fire, fueled his forges beneath the island to manufacture weapons for the other gods. Geologic study suggests, instead, that eruptions take place when hot magma, formed by melting inside the Earth, rises through the crust and emerges at the surface. No one believes the Roman myth anymore, but the island's name evolved into the English word **volcano**, which geologists use to designate either an erupting vent through which molten rock reaches the Earth's surface or a mountain built from the products of eruption.

On the main peninsula of Italy, not far from Vulcano, another volcano, Mt. Vesuvius, towers over the Bay of Naples. Two thousand years ago, a prosperous Roman resort and trading town

FIGURE 5.1 The eruption of Vesuvius devastated and buried Pompeii and nearby Herculaneum in 79 C.E.

(a) In this 1817 painting, the British artist J.M.W. Turner depicted the cataclysmic explosion.

(b) Excavations exposed the ruins of Pompeii. Vesuvius towers behind; dashed lines give its profile before eruption.

(c) A plaster cast of a Pompeii resident who was buried in ash.

named Pompeii sprawled at the foot of Vesuvius. One morning in 79 C.E., earthquakes signaled the mountain's awakening. At 1:00 P.M. on August 24, a dark mottled cloud, streaked by lightning, boiled up above Mt. Vesuvius's summit to a height of 27 km. The cloud spread over Pompeii and turned day into night. Blocks and pellets of rock fell like hail, while fine ash and choking fumes filled the air (**Fig. 5.1a, b**). Frantic people rushed to escape, but for many it was too late. As the growing weight of volcanic debris began to crush buildings, a scalding, turbulent current of ash mixed with pumice fragments surged down the flank of the volcano and swept into Pompeii. By the next day, the town had vanished beneath a 6-m-thick gray-black blanket. This covering protected the ruins of Pompeii so well that when archaeologists excavated the town 1,800 years later, they found an amazingly complete record of Roman daily life. In addition, they discovered open spaces in the debris covering Pompeii. Out of curiosity, they filled the spaces with plaster and then dug away the surrounding ash. The spaces turned out to be fossil casts of Pompeii's unfortunate inhabitants, their bodies forever twisted in agony or huddled in despair (**Fig. 5.1c**).

Clearly, volcanoes are unpredictable and dangerous. Volcanic activity can build a towering, snow-crested mountain or can blast one apart. It can provide the fertile soil and mineral deposits that enable a civilization to thrive, or a rain of destruction that can snuff one out. Because of the diversity of volcanic activity and its consequences, this chapter sets out ambitious goals. We first look more closely at the products of volcanic eruptions and the basic characteristics of volcanoes. Then we consider the different kinds of volcanic eruptions on Earth and why they occur where they do. Finally, we consider the hazards posed by volcanoes, efforts by geoscientists to predict eruptions and help minimize the damage they cause, and the possible influence of eruptions on climate and civilization.

5.2 The Products of Volcanic Eruptions

The drama of a volcanic eruption transfers materials from inside the Earth to our planet's surface. Products of an eruption come in three forms—lava flows, pyroclastic debris, and gas. Note that we use the name *flow* for both a molten, moving layer of lava and for the solid layer of rock that forms when the lava freezes.

Lava Flows

Sometimes it races down the side of a volcano like a fast-moving, incandescent stream, sometimes it builds into a rubble-covered mound at a volcano's summit, and sometimes it oozes like a sticky but scalding paste. Clearly, not all lava behaves in the same way when it rises out of a volcano. Therefore, not all lava flows look the same. Why? The character of a lava primarily reflects its *viscosity* (resistance to flow), and not all lavas have the same viscosity. Differences in viscosity depend, in turn, on chemical composition, temperature, gas content, and crystal content. Silica content plays a particularly key role in controlling viscosity. As noted in Chapter 4, silica-poor (basaltic) lava is less viscous, and thus flows farther than does silica-rich (rhyolitic) lava (**Fig. 5.2**). To illustrate the different ways in which lava behaves, we now examine flows of different compositions.

Basaltic lava flows. Basaltic (mafic) lava has very low viscosity when it first emerges from a volcano because it contains relatively little silica and is very hot. Thus, on the steep slopes near the summit of a volcano, it can flow very quickly, sometimes at speeds of over 30 km per hour (**Fig. 5.3a**). The lava slows down to less-than-walking pace after it starts to cool (**Fig. 5.3b**). Most flows measure less than a few km long, but some flows reach as far as 600 km from the source. How can lava travel such distances? Although all the lava in a flow moves when it first emerges, rapid cooling causes the surface of the flow to crust over after the flow has moved a short distance from the source. The solid crust serves as insulation, allowing the hot interior of the flow to remain liquid and continue to move. As time progresses, part of the flow's interior solidifies, so eventually, molten lava moves only through a tunnel-like passageway, or **lava tube**, within the flow—the largest of these may be tens of meters in diameter. In some cases, lava tubes drain and eventually become empty tunnels.

The surface texture of a basaltic lava flow when it finally freezes reflects the timing of freezing relative to its movement. Basalt flows with warm, pasty surfaces wrinkle into smooth, glassy, rope-like ridges; geologists have adopted the Hawaiian word **pahoehoe** (pronounced "pa-hoy-hoy") for such flows (**Fig. 5.3c**). If the surface layer of the lava freezes and then breaks up due to the continued movement of lava underneath, it becomes a jumble of sharp, angular fragments, creating a rubbly flow also called by its Hawaiian name, **a'a'** (pronounced "ah-ah") (**Fig. 5.3d**). Footpaths made by people living in basaltic volcanic regions follow the smooth surface of pahoehoe rather than the foot-slashing surface of a'a'.

During the final stages of cooling, lava flows contract, because rock shrinks as it loses heat, and may fracture into polygonal columns. This type of fracturing is called **columnar jointing** (**Fig. 5.3e**).

Basaltic flows that erupt underwater look different from those that erupt on land because the lava cools so much more quickly in water. Because of rapid cooling, submarine basaltic lava can travel only a short distance before its surface freezes, producing a glass-encrusted blob, or "pillow" (**Fig. 5.3f**). The rind of a pillow momentarily stops the flow's advance, but within minutes the pressure of the lava squeezing into the pillow breaks the rind, and a new blob of lava squirts out, freezes, and produces another pillow. In some cases, successive pillows add to the end of previous ones, forming worm-like chains.

Andesitic and rhyolitic lava flows. Because of its higher silica content and thus its greater viscosity, andesitic lava cannot flow as easily as basaltic lava. When erupted, andesitic lava first forms a large mound above the vent. This mound then advances slowly down the volcano's flank at only about 1 to 5 m a day, in a lumpy flow with a bulbous snout. Typically, andesitic flows are less than a few km long. Because the lava moves so slowly, the outside of the flow has time to solidify; so as it moves, the surface breaks up into angular blocks, and the whole flow looks like a jumble of rubble called **blocky lava**.

Rhyolitic lava is the most viscous of all lavas because it is the most silicic and the coolest. Therefore, it tends to accumulate either above the vent in a lava dome (**Fig. 5.4**), or in short and bulbous flows rarely more than 1 to 2 km long. Sometimes rhyolitic lava freezes while still in the vent and then pushes upward as a column-like spire up to 100 m above the vent. Rhyolitic flows, where they do form, have broken and blocky surfaces.

Volcaniclastic Deposits

On a mild day in February 1943, as Dionisio Pulido prepared to sow the fertile soil of his field 330 km (200 miles) west of Mexico City, an earthquake jolted the ground, as it had dozens of times in the previous days. But this time, to

FIGURE 5.2 The character of a lava flow depends on its viscosity.

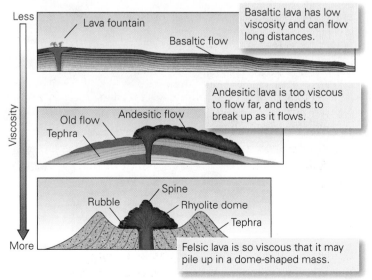

Basaltic lava has low viscosity and can flow long distances.

Andesitic lava is too viscous to flow far, and tends to break up as it flows.

Felsic lava is so viscous that it may pile up in a dome-shaped mass.

FIGURE 5.3 Features of basaltic lava flows. They have low viscosity and thus can flow long distances. Their surface and interior can be complex.

(a) A fast-moving flow coming from Mt. Etna, Sicily.

(b) A basaltic lava flow covers a highway in Hawaii.

(c) Pahoehoe from a recent lava flow in Hawaii. Note the coin for scale.

(d) The rubbly surface of an a'a' flow, Sunset Crater, Arizona.

(e) Columnar jointing develops when the interior of a flow cools and cracks. Such jointing develops in dikes and sills. This example is Devils Postpile in California.

(f) Pillow basalt develops when lava erupts underwater. Later uplift may expose pillows above sea level, as in this Oregon outcrop.

Dionisio's amazement, the surface of his field visibly bulged upward by a few meters and then cracked. Ash and sulfurous fumes filled the air, and Dionisio fled. When he returned the following morning, his field lay buried beneath a 40-m-high mound of gray cinders—Dionisio had witnessed the birth of Paricutín, a new volcano. During the next several months, Paricutín erupted continuously, at times blasting clots of lava into the sky like fireworks. By the following year, it had become a steep-sided cone 330 m high. Nine years later, when the volcano ceased erupting, its lava and debris covered 25 square km.

This description of Paricutín's eruption, and that of Vesuvius at the beginning of this chapter, emphasizes that volcanoes can erupt large quantities of fragmental igneous material. Geologists use the general term **volcaniclastic deposits** for accumulations of this material. Volcaniclastic deposits include **pyroclastic debris** (from the Greek *pyro*, meaning fire), which forms from lava that flies into the air and freezes. They also include the debris formed when an eruption blasts apart preexisting volcanic rock that surrounds the volcano's vent, the debris that accumulates after tumbling down the volcano in landslides or after being transported in water-rich slurries, and the debris formed as lava flows break up or shatter. Let's look at these components in more detail.

> **Did you ever wonder . . .**
> has anyone ever seen a brand-new volcano form?

Pyroclastic debris from basaltic eruptions. Basaltic magma rising in a volcano may contain dissolved volatiles (such as water). As such magma approaches the surface, the volatiles form bubbles. When the bubbles reach the surface, they burst and eject clots and drops of molten magma upward to form dramatic fountains (**Fig. 5.5a**). To picture this process, think of the droplets that spray from a newly opened bottle of soda. Solidification of the pea-sized fragments of glassy lava and scoria produces a type of **lapilli** (from the Latin word for little stones). Pieces of this type of lapilli are informally known as cinders. Rarely, flying droplets may trail thin strands of lava, which freeze into filaments of glass known as Pelé's hair, after the Hawaiian goddess of volcanoes, and the droplets themselves freeze into tiny streamlined glassy beads known as Pelé's tears. Apple- to refrigerator-sized fragments called **blocks** (**Fig. 5.5b**) may consist of already-solid volcanic rock, broken up during the eruption—such blocks tend to be angular and chunky. In some cases, however, blocks form when soft lava squirts out of the vent and then solidifies—such blocks, also known as **bombs**, have streaked, polished surfaces.

Pyroclastic debris from andesitic or rhyolitic eruptions. Andesitic or rhyolitic lava is more viscous than basalt, and may be more gas-rich. The lava flows tend to be blocky to start with, and blocks of flows may tumble down the volcano. Eruptions of these lavas also tend to be explosive. Debris ejected from explosive eruptions includes fragments of pumice and ash. **Ash** consists of particles less than 2 mm in diameter, made from both glass shards formed when frothy lava explosively breaks up during an eruption, and from pulverized pre-existing volcanic rock (**Fig. 5.6a**). Two types of lapilli are produced by explosive eruptions: **pumice lapilli** consists of angular pumice fragments formed from frothy lava (**Fig. 5.6b**); **accretionary lapilli** consists of snowball-like lumps of ash formed when ash mixes with water in the air and then sticks together (**Fig. 5.6c**).

Much of the pyroclastic debris erupted from an exploding volcano billows upward in a turbulent cloud that can reach stratospheric heights (**Fig. 5.6d**). Some, however, rushes down the flank of the volcano in an avalanche-like current known as a **pyroclastic flow** (**Fig. 5.6e**). Pyroclastic flows were once known as *nuées ardentes* (French for glowing cloud), because the debris they contain can be quite hot—200°C to 450°C.

Unconsolidated deposits of pyroclastic grains, regardless of size, constitute **tephra**. Ash, or ash mixed with lapilli, becomes **tuff** when buried and transformed into coherent rock. Tuff that formed from ash and/or pumice lapilli that fell like snow from the sky is called *air-fall tuff*, whereas a sheet of tuff that formed from a pyroclastic flow is an **ignimbrite**. Ash and pumice lapilli in an ignimbrite is sometimes so hot that it welds together to form a hard mass.

Other volcaniclastic deposits. In cases where volcanoes are covered with snow and ice, or are drenched with rain, water mixes with debris to form a **volcanic debris flow** that moves downslope like wet concrete. Very wet, ash-rich debris flows

FIGURE 5.4 This rhyolite dome formed about 650 years ago, in Panum Crater, California. Tephra (cinders) accumulated around the vent.

Tephra cone

Rhyolite dome

FIGURE 5.5 Pyroclastic debris from basaltic eruptions.

(a) Fountains of lava may erupt from basaltic volcanoes.

Blocks and bombs litter the slope of a Hawaiian volcano.

Bombs have smooth, streaked surfaces.

(b) Apple-sized or larger chunks of rock blasted out of a volcano are called blocks. If the lava is soft when it squirts out, it forms streamlined bombs.

become a slurry called a **lahar**, which can reach speeds of 50 km per hour and may travel for tens of kilometers. When debris flows and lahars stop moving, they yield a layer consisting of volcanic debris suspended in ashy mud.

Volcanic Gas

Most magma contains dissolved gases, including water, carbon dioxide, sulfur dioxide, and hydrogen sulfide (H_2O, CO_2, SO_2, and H_2S). In fact, up to 9% of a magma may consist of gaseous components, and generally, lavas with more silica contain a greater proportion of gas. Volcanic gases come out of solution when the magma approaches the Earth's surface and pressure decreases, just as bubbles come out of solution in a soda when you pop the bottle top off.

In low-viscosity magma, gas bubbles can rise faster than the magma moves, and thus most reach the surface of the magma and enter the atmosphere before the lava does. Thus some volcanoes may, for a while, produce large quantities of steam, without much lava (**Fig. 5.7a**). The last bubbles to form, however, freeze into the lava and become holes called **vesicles** (**Fig. 5.7b**). In high-viscosity magmas, the gas has trouble escaping because bubbles can't push through the sticky

lava. When this happens, explosive pressures build inside or beneath the volcano.

> ## Take-Home Message
>
> Volcanoes erupt lava, pyroclastic debris, and gas. The character of a lava flow—whether it has low viscosity and spreads over a large area, or has high viscosity and builds a mound over the vent, depends largely on its composition. Pyroclastic debris includes pumice, lapilli, blocks, and bombs. Some may fall over the countryside like snow, but some surges down the flank of a volcano as a pyroclastic flow.

5.3 The Structure and Eruptive Style of Volcanoes

Volcanic Architecture

As we saw in Chapter 4, melting in the upper mantle and lower crust produces magma, which rises into the upper crust.

FIGURE 5.6 The components of an explosive eruption.

(a) Ash flakes

(b) Pumice lapilli, Mexico.

(c) Accretionary lapilli

(d) The 1989–1990 eruption of Redoubt Volcano in Alaska produced a giant cloud of ash that mushroomed up to the stratosphere.

(e) Ash erupts from Mt. Merapi, Indonesia, in 2006. Some of the ash rises into the sky, whereas some rushes down the volcano's flank as a pyroclastic flow.

FIGURE 5.7 The gas component of volcanic eruptions.

(a) A volcano in Alaska erupting large quantities of steam.

(b) Gas bubbles frozen in lava produce vesicles, as in this block from Sunset Crater, Arizona.

Typically, this magma accumulates underground in a **magma chamber**, a zone of open spaces and/or fractured rock that can contain a large quantity of magma. A portion of the magma may solidify in the magma chamber and transform into intrusive igneous rock, whereas the rest rises through an opening, or conduit, to the Earth's surface and erupts from a volcano. The conduit may have the shape of a vertical pipe, or chimney, or may be a crack called a **fissure** (**Fig. 5.8a, b**). At the top of a volcanic edifice, a circular depression called a **crater** (shaped like a bowl, up to 500 m across and 200 m deep) may develop. Craters form either during eruption as material accumulates around the summit vent, or just after eruption as the summit collapses into the drained conduit.

During major eruptions, the sudden draining of a magma chamber produces a **caldera**, a big circular depression up to thousands of meters across and up to several hundred meters deep. Typically, a caldera has steep walls and a fairly flat floor and may be partially filled with ash (**Fig. 5.9a–d**).

Geologists distinguish among several different shapes of subaerial (above sea level) volcanic edifices. **Shield volcanoes**, broad, gentle domes, are so named because they resemble a soldier's shield lying on the ground (**Fig. 5.10a**). They form when the products of eruption have low viscosity and thus are weak, so they cannot pile up around the vent but rather spread out over large areas. **Scoria cones** (informally called **cinder cones**) consist of cone-shaped piles of basaltic lapilli and blocks, generally from a single eruption (**Fig. 5.10b**). **Stratovolcanoes**, also known as composite volcanoes, are large and cone-shaped, generally with steeper slopes near the summit, and consist of interleaved layers of lava, tephra, and volcaniclastic debris (**Fig. 5.10c**). Their shape, exemplified by Japan's Mt. Fuji, supplies the classic image that most people have of a volcano; the prefix *strato-* emphasizes that they can grow to be kilometers high.

Concept of Eruptive Style: Will It Flow, or Will It Blow?

Kilauea, a volcano on Hawaii, produces rivers of lava that cascade down the volcano's flanks. Mt. St. Helens, a volcano near the Washington–Oregon border, exploded catastrophically in 1980 and blanketed the surrounding countryside with tephra. Clearly, different volcanoes erupt differently and, as we've noted, successive eruptions from the same stratovolcano may differ markedly in character from one another. Geologists refer to the character of an eruption as **eruptive style**. Below, we describe several distinct eruptive styles and explore why the differences occur.

Effusive eruptions. The term *effusive* comes from the Latin word for pour out, and indeed that's what happens during

FIGURE 5.8 Crater eruptions and fissure eruptions come from conduits of different shapes.

(a) At a crater eruption, lava spouts from a chimney-shaped conduit.

(b) At a fissure eruption, lava erupts as a curtain along a crack.

an **effusive eruption**—lava pours out a summit vent or fissure, filling a lava lake around the crater and/or flowing in molten rivers for great distances (**Fig. 5.11a**). Effusive eruptions occur where the magma feeding the volcano is hot and mafic and, therefore, has low viscosity. Pressure, applied to the magma chamber by the weight of overlying rock, squeezes magma upward and out of the vent; in some cases, the pressure is great enough to drive the magma up into a fountain over the vent.

Explosive eruptions. When pressure builds in a volcano, the eruption will likely yield an explosion. Smaller explosions take place during basaltic eruptions, when gas builds up and suddenly escapes, spattering lava drops and blobs upward—these then solidify and fall as tephra. Occasionally, a volcano blows up in a huge explosion. Such catastrophic explosions can be triggered by many causes. For example, if a crack forms in the flank of an island volcano, water will enter the magma chamber and suddenly turn to steam, the expansion of which blasts the volcano apart. Such explosions can also happen in felsic or andesitic volcanoes, if very viscous magma plugs the vent until huge pressure builds inside. If the plug eventually cracks, or the flank of the volcano cracks, the gas inside the volcano suddenly expands, and like a giant shotgun blast, it sprays out the molten contents of the volcano and may cause the volcano itself to break apart. Such explosions, awesome in their power and catastrophic in their consequences, can eject cubic kilometers of debris outward. In some cases, the sudden draining of the magma chamber, and the ejection of debris, causes the remnants of the volcano to collapse and form a caldera.

During a large explosion, the force of the blast shoots debris skyward in a vertical column (**Fig. 5.11b**). But the force can only take the material so high. The huge plumes of ash that rise to stratospheric heights above large explosions do so by becoming turbulent, billowing, *convective* clouds. This means that the warm mixture of volcanic ash, gas, and air is less dense than the surrounding, cooler air, so the warm mixture rises buoyantly. The resulting plume resembles a mushroom cloud above a nuclear explosion. Coarser-grained ash and lapilli settle from the cloud close to the volcano, whereas finer ash gets carried farther away. Some ash enters high-elevation winds and will be carried around the globe. The denser components collapse downward once they run out of explosive energy, and gravity pulls them back down. This phenomenon, the "collapse" of the column, produces the pyroclastic flows that surge down a volcano's flanks.

What is a pyroclastic flow like? In 1902, the people of St. Pierre, a town on the Caribbean island of Martinique, sadly found out. St. Pierre was a busy port town, about 7 km south of the peak of Mt. Pelée, a volcano. When the volcano began emitting steam and lapilli, residents of the town became nervous and debated about the need to evacuate. Meanwhile, a rhyolite dome grew and obstructed the throat

FIGURE 5.9 The formation of volcanic calderas.

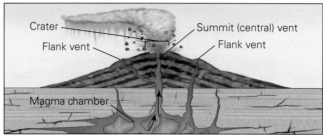

(a) As an eruption begins, the magma chamber inflates with magma. There can be a central vent and one or more flank vents.

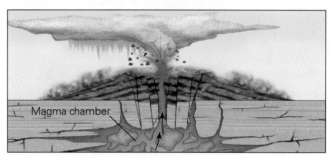

(b) During an eruption, the magma chamber drains, and the central portion of the volcano collapses downward.

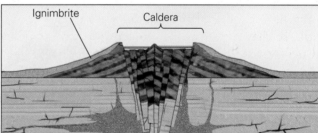

(c) The collapsed area becomes a caldera. Later, a new volcano may begin to grow within the caldera.

Time

(d) This caldera in Oregon formed about 7,700 years ago. Afterward, it filled with water to become Crater Lake. Wizard Island, protruding from the lake, is a cinder cone that grew on top of the caldera floor.

FIGURE 5.10 Different shapes of volcanoes.

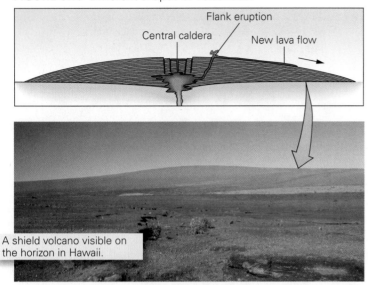

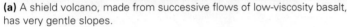

(a) A shield volcano, made from successive flows of low-viscosity basalt, has very gentle slopes.

A shield volcano visible on the horizon in Hawaii.

(b) A cinder cone on the flank of a larger volcano in Arizona. The pile of cinders has assumed the angle of repose. A lava flow covers the land surface in the distance.

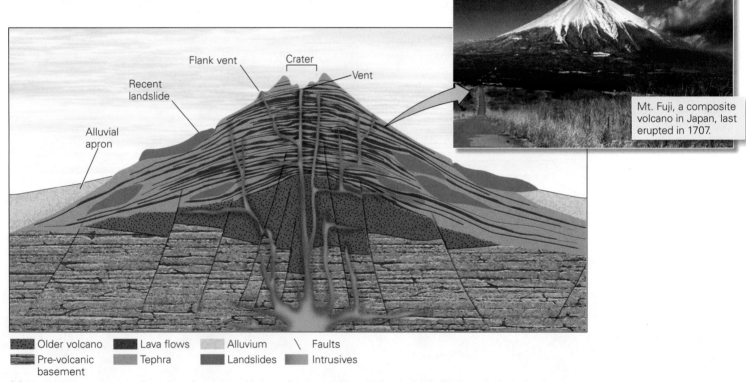

Mt. Fuji, a composite volcano in Japan, last erupted in 1707.

Legend:
- Older volcano
- Pre-volcanic basement
- Lava flows
- Tephra
- Alluvium
- Landslides
- Faults
- Intrusives

(c) A stratovolcano (composite volcano) consists of layers of tephra and lava. Volcanic debris flows and ash avalanches modify slopes and contribute to the development of a classic cone-like shape.

of the volcano. On May 8, the dome suddenly cracked, and the immense pressure that had been building beneath the obstruction was released. In the same way that champagne bursts out of a bottle when you pull out the cork, a cloud of hot ash and pumice lapilli spewed out of Mt. Pelée, and a pyroclastic flow swept down Pelée's flank. Partly riding on a cushion of air, this flow reached speeds of 300 km per hour, and slammed into St. Pierre. Within moments, all the town's buildings had been flattened and all but two of its 28,000 inhabitants were dead of incineration or asphyxiation. Similar eruptions have happened more recently on

FIGURE 5.11 Contrasting eruptive styles.

(a) Effusive (lava-dominated) eruption of Mount Etna in Sicily, Italy, 1992.

The ash built a fan offshore.

(c) The aftermath of a pyroclastic flow down the flank of the Soufrière Hills Volcano, Montserrat (Caribbean).

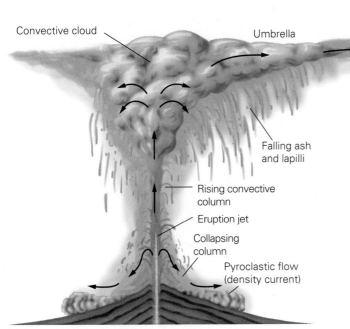

Convective cloud
Umbrella
Wind
Stratospheric haze
Falling ash and lapilli
Rising convective column
Eruption jet
Collapsing column
Pyroclastic flow (density current)

(b) Cross section of an explosive eruption.

eruptions due to fountaining basaltic lava yield cinder cones, and those that alternate between effusive and large pyroclastic eruptions become composite volcanoes (stratovolcanoes). Large explosions yield calderas and blanket the surrounding countryside with ash and/or ignimbrites.

Why are there such contrasts in eruptive style? Eruptive style depends on the viscosity and gas contents of the magma in the volcano. These characteristics, in turn, depend on the composition and temperature of the magma and on the environment (subaerial or submarine) in which the eruption occurs. Traditionally, geologists have classified volcanoes according to their eruptive style, each style named after a well-known example (**Geology at a Glance**, pp. 132–133).

the nearby island of Montserrat, but with a much smaller death toll because of timely evacuation (**Fig. 5.11c**).

Relation of eruptive style to volcanic type. Note that the type of volcano (shield, cinder cone, or composite) depends on its eruptive style. Volcanoes that have only effusive eruptions become shield volcanoes, those that generate small pyroclastic

Take-Home **Message**

At a volcano, lava rises from a magma chamber and erupts from chimney-like conduits or from crack-like fissures. Low-viscosity basalt lava flows build shield volcanoes. Fountaining basalt spatters tephra to build cones. Successive eruptions of pyroclastic debris and lava build stratovolcanoes, and explosions produce calderas.

BOX 5.1   CONSIDER THIS ...

Volcanic Explosions to Remember

Explosions of volcanoes generate enduring images of destruction. The historical record shows a vast range in the volume of debris erupted, even though the largest *observed* eruption (Tambora in 1815) was small compared to a super-explosion that took place over 600,000 years ago in what is now Yellowstone National Park, Wyoming (**Fig. Bx5.1a**). Let's look at two notable examples of explosions.

Mt. St. Helens, a snow-crested stratovolcano in the Cascades of the northwestern United States, had not erupted since 1857. However, geologic evidence suggested that the mountain had a violent past, punctuated by many explosive eruptions. On March 20, 1980, an earthquake announced that the volcano was awakening once again. A week later, a crater 80 m in diameter burst open at the summit and began emitting gas and pyroclastic debris. Geologists who set up monitoring stations to observe the volcano noted that its north side was beginning to bulge markedly, suggesting that the volcano was filling with magma, making the volcano expand like a balloon. Their concern that an eruption was imminent led local authorities to evacuate people in the area.

The climactic eruption came suddenly. At 8:32 A.M. on May 18, a geologist, David Johnston, monitoring the volcano from a distance of 10 km, shouted over his two-way radio, "Vancouver, Vancouver, this is it!" An earthquake had triggered a huge landslide that caused 3 cubic km of the volcano's weakened north side to slide away. The sudden landslide released pressure on the magma in the volcano, causing a sudden and violent expansion of gases that blasted through the side of the volcano (**Fig. Bx5.1b**). Rock, steam, and ash screamed north at the speed of sound and flattened a forest and everything in it over an area of 600 square km (**Fig. Bx5.1c**). Tragically, Johnston, along with 60 others, vanished forever.

Seconds after the sideways blast, a vertical column carried about 540 million tons

FIGURE Bx5.1 Examples of explosive eruptions.

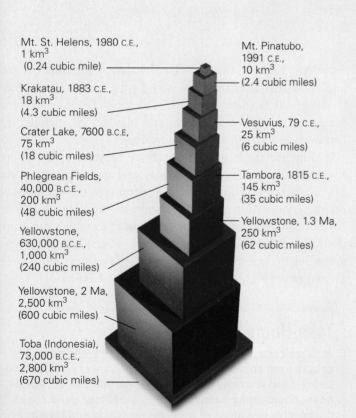

Mt. St. Helens, 1980 C.E.,
1 km³
(0.24 cubic mile)

Krakatau, 1883 C.E.,
18 km³
(4.3 cubic miles)

Crater Lake, 7600 B.C.E,
75 km³
(18 cubic miles)

Phlegrean Fields,
40,000 B.C.E.,
200 km³
(48 cubic miles)

Yellowstone,
630,000 B.C.E.,
1,000 km³
(240 cubic miles)

Yellowstone, 2 Ma,
2,500 km³
(600 cubic miles)

Toba (Indonesia),
73,000 B.C.E.,
2,800 km³
(670 cubic miles)

Mt. Pinatubo,
1991 C.E.,
10 km³
(2.4 cubic miles)

Vesuvius, 79 C.E.,
25 km³
(6 cubic miles)

Tambora, 1815 C.E.,
145 km³
(35 cubic miles)

Yellowstone, 1.3 Ma,
250 km³
(62 cubic miles)

(a) The relative amounts of pyroclastic debris (in cubic km) ejected during major explosive eruptions.

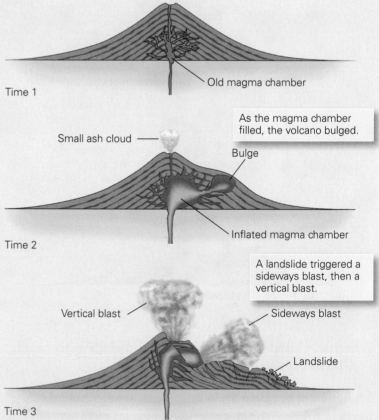

Time 1

Old magma chamber

Small ash cloud

As the magma chamber filled, the volcano bulged.

Bulge

Inflated magma chamber

Time 2

A landslide triggered a sideways blast, then a vertical blast.

Vertical blast

Sideways blast

Landslide

Time 3

(b) Stages during the eruption of Mt. St. Helens, 1980.